DEATH
BEFORE
DISHONOUR

NICHOLAS DAVIES

DEATH BEFORE DISHONOUR

TRUE STORIES OF THE SPECIAL FORCES HEROES WHO FIGHT GLOBAL TERROR

JB

JOHN BLAKE

Published by John Blake Publishing Ltd,
3 Bramber Court, 2 Bramber Road,
London W14 9PB, England

www.johnblakepublishing.co.uk

www.facebook.com/Johnblakepub facebook
twitter.com/johnblakepub twitter

First published in paperback in 2003
This edition published in 2013

ISBN: 9781857826777

British Library Cataloguing-in-Publication Data:

A catalogue record for this book is available from the British Library.

Design by www.envydesign.co.uk

Printed in Great Britain by CPI Group (UK) Ltd

1 3 5 7 9 10 8 6 4 2

Papers used by John Blake Publishing are natural, recyclable products made from
wood grown in sustainable forests. The manufacturing processes conform to the
environmental regulations of the country of origin.

Every attempt has been made to contact the relevant copyright-holders, but some were
unobtainable. We would be grateful if the appropriate people could contact us.

CONTENTS

The Ten Commandments for Special Warfare by US Navy SEAL Richard Marcinko, 1970

1. I am the War Lord and the wrathful god of Combat and I will always lead you from the front.

2. I will treat you all alike – and like shit.

3. Thou shalt do nothing I shall not do first.

4. I shall punish thy bodies for the more thou sweatest in training, the less thou bleedest in combat.

5. Verily, if thou hurteth in thy efforts and suffer pain, thou are Doing It Right.

6. Thou hast not to like it – thou hast just to do it.

7. Thou shalt keep it simple, stupid.

8. Thou shalt never assume anything, thou wilt check it out.

9. Verily thou art paid not for thy methods but for thy results and thou wilt kill thy enemy before he killeth you, and by any means available.

10. Thou shalt bear in mind the Ultimate Commandment: There are no rules; thou shalt win at any price.

FOREWORD

A book of this nature, revealing the inside story of so many missions undertaken by the world's foremost Special Forces, could not have been written without a great deal of assistance, advice and expert knowledge from many people. Many of the operational details were kindly provided by those who took part but who were bound by allegiance to their colleagues not to reveal names. I readily agreed to abide by this code, for I wanted in no way to put them at risk.

I was also given great assistance by a number of defence officials and Special Force advisers working in London for governments around the world. They kindly furnished me with much of the detail of the operations in which their respective forces had been involved, and went to considerable trouble to find out, from their governments back home, the information I had requested.

In the Bibliography I have listed the many sources in which I was able to cross-check the facts of many of the operations. I owe a debt of gratitude to all of the authors.

I chose deliberately to write about some of the unsuccessful operations, even disasters, describing operational errors which too often put the lives of Special Forces soldiers in danger. Indeed in some instances men lost their lives as a result of such mistakes.

In consequence this is not simply another gung-ho epic underlining the heroism of the world's elite forces but, I hope, a balanced account of many of the extremely demanding and dangerous circumstances in which such soldiers have found themselves. To them all, we owe more than we realise.

Nicholas Davies,
September 2002

CHAPTER 1

THE SPECIAL SOLDIER

SOLDIERS SPECIALLY TRAINED to complement the use of regular armed forces are not a new phenomenon. Military historians have come to realise that over hundreds and even thousands of years groups of highly skilled fighting men have carried out the same type of operations as are entrusted to today's Special Forces. The Bible recounts how King David's special soldiers carried out surprise night raids against the Philistines. In the twelfth century the Mongol leader Genghis Khan often sent bands of irregular horsemen on covert missions behind enemy lines.

During the past thirty or forty years, in response to the spread in many countries of terrorism, sedition and underground activities by disaffected groups, the tasks of the modern special soldier have become more demanding, yet his function has remained remarkably constant. Today's Special Forces soldiers understand that the only 'special' part of their role are the particularly demanding tasks that their unit is ordered to carry out, the training necessary to undertake these and the character of the men performing them.

Throughout the world these soldiers are defined by the role they perform and the training they receive. Almost always they operate in small groups, frequently under cover of darkness and often behind enemy lines. They usually rely on the latest technology for both communication and subversive action. The Special Forces soldier therefore has to be skilled in the latest military techniques, which he uses mainly in reconnaissance, raiding missions and other irregular operations.

But Special Forces commanders agree that what is important above all else about this breed of soldier is the character, the quality of the man. Many of them believe that such soldiers are adventurers at heart, men who, having volunteered to take part in daring, front-line actions, can face the prospect of death or injury with equanimity. They look for men who have certain basic qualities – courage, intelligence, tenacity, common sense and, indispensably, a sense of humour.

Colonel Charlie Beckwith, the founder of America's Delta Force, put it this way, 'I had to articulate that before a soldier could become a good unconventional soldier he'd first have to be a good conventional soldier. He had to understand what a rifle squad was all about, what a platoon could do, what a rifle company needed to know. You can't be unconventional until you are conventional first.'

Basic advice to American Special Forces soldiers was laid down in 1759 and remains much the same today. Indeed the original list of instructions is still displayed in some barracks of the US Rangers:

1. **Don't forget nothing.**

2. **Have your musket as clean as a whistle, hatchet scoured and sharp, sixty rounds of ball ready. Be ready to move at a minute's notice.**

3. **When on the march, act as if you are hunting a deer. Get your shot in first.**

4. Tell the truth on what you see and do. The army depends on us for information. Never lie to a Ranger officer.

5. Don't take any chance you don't have to.

6. When on the march go single file, far enough apart that one shot won't kill two men.

7. On soft ground, spread out so it's hard to track us.

8. When we march we keep moving until dark, so the enemy can't get at us.

9. When we camp, half sleep while the other half watch.

10. Don't ever march home the same way; take a different route to avoid ambush.

11. In big parties or little ones, keep a scout twenty yards ahead or behind and to either flank, so we won't be surprised.

12. Every night you will be told what to do if attacked by a superior force.

13. Don't sit down to eat without posting sentries.

14. Don't sleep beyond dawn; that's when the enemy likes to attack.

15. Don't cross rivers by the regular ford.

16. Ambush the folks that are trying to ambush you.

17. Don't stand when the enemy is coming against you; kneel down, lie down, get behind a tree.

18. Let the enemy come close, close enough to touch. Then let him have it and finish him off with your hatchet.

Even today most of the world's Special Forces, including Britain's Special Air Service, or SAS, and the US Rangers, make use of this wisdom in their operations.

The history of the past few hundred years is dotted with missions undertaken by special soldiers using these tactics, which demonstrates that ambushing and skirmishing can prove highly successful against a large regular army. For example, in the Peninsular War of the early nineteenth century, the Duke of Wellington realised his British forces were not sufficient to tackle Napoleon's French army and used Spanish irregular forces to harass the enemy. In Spanish, those men were called *guerrillas* and the name stuck.

A few years later, during Napoleon's famous retreat from Moscow in 1812, it was Cossack skirmishers who harried and ambushed the great French army as it headed back home through the snow, inflicting far more damage than the regular Russian forces had.

In the American Civil War of 1861–5 both Union and Confederate generals employed skirmishing tactics and used guerrillas behind enemy lines to cause maximum damage with little loss of life. In the Boer War of 1899–1902 the South African Boers used 'commandos' – small units of horsemen – to harass the might of the British Army to great effect.

During the Great War of 1914–18, when the massive French and British armies repeatedly failed to penetrate the enemy network of defences in massive frontal assaults, Allied Special Forces played a very limited role in the Allied victory. It was sustained deployment of regular forces, with a huge loss of life on both sides, that eventually defeated the Germans.

The idea of Special Forces remained very much alive, however, not least in Germany. In 1935, two years after Hitler had come to power, a naval captain, Wilhelm Canaris, was appointed Chief of the Abwehr, or Counter-Intelligence. Over the following years, with the prospect of another major war certain, Canaris established a large Intelligence apparatus which included a section responsible for special military units and sabotage. It was within this section that Germany's first Special Force, the Brandenburgers, came into being on the eve of World War Two. In planning the role of these soldiers, military advisers studied the exploits of Colonel T.E. Lawrence – Lawrence of Arabia – during the previous war. They examined how Lawrence and a small, elusive band of saboteurs had been able to create chaos among the Turkish enemy in Arabia, bring confusion to the Turkish forces and win victories out of all proportion to their numbers even though they had no armour, no heavy weapons, no guns and no back-up.

Canaris decided that the new force would consist of small, highly mobile units and that this mobility would be the key to defeating a much larger enemy. He now needed to select the right type of men for the job. Recruits should be men who had lived overseas and been engaged on the land or in open-air activities, where they would have become tough, independent and strong-willed and gained a knowledge of foreign languages, cultures and customs. Naturally these young men should be super-fit, but Canaris further decreed that they should all be volunteers. This rule of accepting only volunteers into the Brandenburg units was to be maintained throughout World War Two.

The first recruits were all racial Germans who lived in German communities outside the borders of the Third Reich. Many lived in small towns and villages close to the German and Austrian borders, where most people spoke German and the language of the neighbouring country, for example, Poland, Hungary or Czechoslovakia. Indeed Canaris, by now promoted to the rank of Admiral, would boast that there was not an area or country in

Europe with which Brandenburgers were not familiar, nor a language they could not speak fluently.

The training was severe. Recruits were taught the usual Special Forces skills, such as parachuting, weapons handling, skiing, marksmanship and the use of small boats and canoes. Their instruction also included fieldcraft, which involved having to survive on food they gathered from the wild. All exercises were undertaken using live ammunition; all recruits were schooled in producing explosives using only flour, icing sugar and potash; and methods of silent killing included the garrotte and the hunting knife. Much of this training was conducted in the coldest weather to hone the volunteers into first-class soldiers and saboteurs capable of spearheading battles, gaining vital objectives and generally seizing the initiative.

There was, however, one major difference between this German formation and most of the other Special Forces that would be created during the next half century. The Brandenburgers carried out many of their operations fully or partly disguised and, in some cases, wearing no item of uniform whatsoever. Full disguise meant that every soldier wore an enemy uniform; partial disguise meant wearing an enemy helmet or trench coat. On some occasions Brandenburgers would wear civilian clothes but carry concealed weapons.

This elite body of fighting men won their spurs in May 1940, during the first vital hours of Hitler's strikes into new territory. At the forefront of the main German attacks were Brandenburg units tasked with seizing and holding four bridges until the huge conventional forces of Panzer (tank) troops and infantry arrived to drive across the bridges without hindrance.

The ground had been prepared thoroughly. Detachments of Brandenburgers fluent in the languages of the countries to be invaded and disguised as local farmers and peasants had spent the previous three months slipping back and forth into enemy territory to report back to their commanders on enemy numbers and defences. By the time of the invasion the German high command had an intimate knowledge of all aspects of the resistance they would face.

The four hundred-metre-long Gennep bridge, near the Dutch town of that name, carried the railway line from Goch in Germany across the River Meuse and into the Netherlands. The Germans had learnt that the bridge had been primed with explosives which could be detonated by one sentry in a matter of seconds. Senior officers believed that if the Dutch were given time to destroy it the first phase of Hitler's Blitzkrieg would be a failure. The German plan was for the Brandenburgers to attack and kill the sentries before the bridge could be blown up. This achieved, two fully laden troop trains would cross it and begin the war on the western front.

At midnight on May 9 a dozen Brandenburgers dressed as Dutch military policemen crossed the Meuse two kilometres upstream of the bridge and made their way to the river bank. No one challenged them. At 1 am they were in concealed positions, and remained there until dawn, when they heard the steam trains making their way towards the bridge. As the Germans approached the bridge, the Dutch sentries on duty ran to intercept them, raising their rifles ready to shoot. But the sentries stopped when they realised the six men facing them on the other side of the bridge were wearing the uniform of the Dutch military police. Within seconds the sentries had been grabbed from behind by the Brandenburgers, who slit their throats. The Germans now controlled the eastern side of the bridge.

On the Dutch side of the bridge, other sentries then heard the train approaching from Germany, and they were under orders to blow up the bridge if any trains approached. On duty in the middle of the bridge – from where the explosives could be detonated – was a lone elderly sentry. He watched spellbound as the train approached and before he realised that he should detonate the charge six men had leapt from the slowly passing train. He was killed and the detonator defused. The invasion of the Netherlands had begun without a single German soldier being killed or wounded. The operation, the first major mission carried out by Germany's Special Forces, had been a complete success. In fact this was only one of the attacks by Brandenburgers in the war's Blitzkrieg phase, during which the new units proved to be highly effective. The daring of these special soldiers had opened the way for the invading German armoured and infantry divisions to make their extraordinary dash to the English Channel, sweeping aside the defences of western European countries within a few weeks.

Just a few weeks after the invasion, a Brandenburg officer, Lieutenant Klaus Grabert, was tasked with a special mission. He was to select twelve of his best men for an audacious raid which would prevent the Belgians opening sluice gates and flooding the entire area around the town of Nieuwpoort, as this would halt the German advance. Exactly the same strategy by the Belgians had thwarted the Germans in the Great War.

Within twenty-four hours the chosen twelve arrived in Ghent for briefing. Their task was to foil the Belgian plan by capturing the pumping station on the south bank of the River Yser. Belgian army greatcoats and caps had been collected and the men were taken to Ostend in a captured Belgian military bus. In the chaos of the ongoing fighting the bus passed unchallenged through thousands of unarmed Belgian troops who clogged the roads and Ostend itself. Once in the city, a French-speaking Brandenburger asked what was going on and was surprised to hear that the Belgians had surrendered and the British forces had dug in at Nieuwpoort and were still fighting. The Germans also learnt that charges had been laid at the bridge into Nieuwpoort.

The Brandenburgers drove on, though the road was so clogged with traffic, fleeing refugees and Belgian troops making their way back to Ostend that the twenty-five-kilometre journey took four hours. But still no one stopped the bus to ask questions. The British garrison at Nieuwpoort was small, consisting of a few Lancers in armoured cars and some infantry platoons. Heading at speed towards Nieuwpoort behind the Brandenburgers was the powerful German XXVI Corps, which had orders to attack the British forces now concentrating on Dunkirk and wipe them out in the town and on the beaches.

As the sun began to sink in the west the bus arrived at the bridge and immediately came under fire from British troops on the other side of the river who were guarding the vital crossing place. The driver brought the vehicle to a skidding halt on the bridge, swinging it broadside so that it formed a barrier. Out leapt the twelve men, who tore off their Belgian greatcoats and began returning fire with machine guns and rifles.

Lieutenant Grabert and a corporal made a plan of action. The two of them would wait until dark and then crawl across the bridge towards the British position. When they came across any wires which they suspected led to the explosive charges they would cut them and keep moving forward. When they finally reached the other side of the bridge they would open fire with their machine pistols, the signal for the other Brandenburgers to storm across the bridge. On reaching the British side the twelve men would then spread out, shout orders and fire weapons from different angles at different targets, so as to give the impression that they were only the forward unit of a much larger force. It was the sort of desperate, some would say suicidal, gamble that many other Special Forces would copy in future years.

Sliding along on their stomachs, Grabert and the corporal each carried insulated wire cutters in one hand and a machine pistol in the other. It wasn't long before they discovered the wires leading to the explosive charges. To their anguish, the charges had been fixed to the structure of the bridge, which meant they would have to crawl along the footpath rather than the road. This, they feared, could put the entire operation at risk, because if any Very lights were fired over the bridge and illuminated them, they would be exposed to British fire

Each time a Very light was fired the two Germans froze, praying that the British machine-gunners would not spot them. Sporadically, the British gunners would lay down some rapid fire and these bullets passed only centimetres above the heads of the two Germans. They continued to move forward on their bellies, but only immediately after a Very light had expired, because then there were a few moments of absolute darkness. They discovered another pair of wires and then, some twenty metres further on, a third pair; they cut both. They were now certain they had made safe the bridge and they put the next part of their bold plan into operation.

Sheltering behind a girder near the British-held end of the bridge, they opened fire with the machine pistols, each firing three magazines at the enemy positions. Grabert also threw three grenades at the machine-gun post. At the sound of their comrades' machine pistols the ten other Brandenburgers leapt to their feet and ran flat out across the bridge, firing as they went.

Sixty seconds later the twelve Germans had formed a group and all began firing at will, hurling hand grenades and causing confusion among the British soldiers. One enemy position after another was taken by storm as the dozen fearless soldiers threw grenades and followed these with rapid machine-gun fire. The tiny group of defenders were soon pushed back from the pump house and three Brandenburgers checked that the sluice gates had not been opened or primed with explosive charges.

Expecting a counter-attack, the twelve men took up defensive positions, but none came. One hour later Grabert and his corporal cautiously moved forward to check the British positions, only to find that the British soldiers had disappeared into the night. Not one Brandenburg man had been killed or seriously wounded and the mission had been a complete success. In a short, sharp mission executed with skill and courage, the twelve Brandenburgers had prevented the Allies from flooding the area along the Flanders coast – and the way was now wide open for the Germans to advance to the Dunkirk beaches some forty kilometres to the south.

Brandenburg units were also used extensively in Hitler's invasion of Russia in 1941. Their tasks were made much easier by the fact that Finland had sold Germany scores of tanks, trucks, uniforms and greatcoats captured during its war with Russia in 1939–40. The German high command had listed a hundred separate targets along the Russian frontier which Brandenburg special soldiers could be tasked to take by any means. This would prepare the way for the mass of German armoured and infantry divisions, which could once again employ the same Blitzkrieg tactics as had proved so successful against the western allies. By capturing key airfields, bridges and road junctions, the Brandenburg detachments would allow the Panzer divisions to roll into Russia that much faster.

Being equipped with Russian military vehicles and uniforms made it easier for the Brandenburgers to operate behind the enemy lines. No questions were asked as they went about their clandestine missions inside Russian territory. Indeed the success of the German invasion of Russia was due in large measure to the myriad of missions the Brandenburgers had carried it in those crucial first few days of the offensive.

But there would be further tasks in quite different theatres of war for the Brandenburg units. One such area was North Africa, although no one in the German high command believed this was an ideal location for their particular skills. A German army under the great General Erwin Rommel had been sent to North Africa to support the broken Italian forces, which had been all but crushed by the British army under General Wavell. So fast had been Wavell's advance against the Italians that most of Mussolini's so-called African Empire was on the verge of defeat. Rommel saw that what was needed was men with knowledge of Africa from families who had lived and worked in the German possessions in East and South West Africa; Germans who could speak Arabic, Swahili and English and who understood the African way of life. Volunteers were invited to join the Brandenburgers' new Afrika Kompanie and within weeks sixty former émigrés had been selected and trained. Rommel wanted the Afrika Kompanie to work behind British lines, reporting back by wireless on the location, size and equipment of the British forces they came across.

Small groups from the Afrika Kompanie carried out such operations in the desert, but because most of them were unable to speak English without a distinct German accent, much of the information they gathered was not particularly crucial or even accurate.

However, in 1942 Rommel believed he was on the verge of crushing the British Eighth Army, and he planned, after achieving this, to drive on through Egypt to the Nile and grab the vital link, the Suez Canal. For this campaign he would need skilful Special Forces men. Rommel called the Brandenburg commanders and outlined the tasks he would want the Afrika Kompanie to carry out. Their first task would be to seize the bridges over the Nile and the Suez Canal to prevent their destruction by the British, and then to hold them until Rommel's Panzer divisions could break through and join up with the Brandenburgers. It was a tough mission but one which never materialised, for General Montgomery would rally his troops, defeat Rommel and his

Panzers in the critical battle of El Alamein and then, some three months later, smash the German defences and drive them out of North Africa.

But now, as he faced Montgomery's Eighth Army, Rommel gave the Afrika Kompanie a new, vitally important but extraordinary task – to locate and trace the route the British were using to supply reinforcements to Montgomery. He had been informed by German Intelligence that the British were landing tanks, guns, small arms, ammunition, spares and other equipment in Nigeria and transporting them across some fifteen hundred miles of rough terrain, as well as the Sahara Desert, to Cairo. Rommel charged the Afrika Kompanie with the task of determining the exact route of this supply line so that it could be harried by German forces and cut.

First, the Afrika Kompanie needed to acquire British vehicles, uniforms and weapons so that any British troops they came across would believe they were members of the British Long Range Desert Group. Most of the route would pass through countries friendly to Britain, so it was necessary for the Germans to portray themselves as British. The Afrika Kompanie also acquired a British Spitfire, which they would use as a long-distance reconnaissance plane as Brandenburgers on the ground made their way from Egypt to Nigeria. The Spitfire, with Royal Air Force markings, would fly several hundred kilometres ahead of the group, circle and return to the Brandenburgers. It was hoped that the aircraft might come across British reinforcements making their way to Cairo and return to the group each day with vital information on such troop movements.

The Afrika Kompanie left Libya with twelve fifteen-hundredweight trucks, twelve half-tracks fitted with two-pounder guns, four jeeps carrying anti-aircraft machine-guns, a staff car, a wireless vehicle, a petrol tanker, a workshop vehicle and a rations vehicle. The column travelled due south to Al Qatrun, some two hundred kilometres from the border with Niger, the country between Libya and Nigeria, and set up their headquarters, which included a communications base and a rough airstrip for the Spitfire. They also left at Al Qatrun two half-tracks equipped with machine guns in case of attack from Arab brigands. They waited four days for the all-important Spitfire to arrive but to no avail. It never turned up, which meant the Afrika Kompanie now had a much more hazardous and difficult task. Apparently, the German aero engineers were unable to get the captured aircraft into the air.

One small group of Afrika Kompanie soldiers drove west from Al Qatrun into Algeria to carry out a recce of the French colony in case supplies for Montgomery's army were being brought through southern Algeria. Another group drove south-east to the Tibesti mountains in northern Chad. A third group, the largest, would search for supply lines in southern Algeria's Tassili mountain range, some six hundred kilometres in length and reaching more than fifteen hundred metres in places.

If any of the three groups failed to discover the Allied supply route, their orders were to continue the search, criss-crossing the arid, desolate wastes of the Sahara desert on foot and in the searing heat of summer. All that they managed to discover was that French forces controlled the two mountain ranges. They had found no supply routes and no evidence of one having existed. Rommel was not impressed. The mission was remarkable, however, because it showed the extraordinary resilience, tenacity and adaptability to exceptional circumstances that tough, well-trained Special Forces could display in the most inhospitable terrain.

The final mission of the Brandenburgers in World War Two was a gallant, heroic battle fought with extraordinary courage despite the utter futility of the orders they had been given. The manner in which those men carried out the orders was a magnificent example of bravery, a quality which has continued to characterise special soldiers to the present day.

At the end of March 1945 the 600 Brandenburg Paratroop Battalion was put into the German bridgehead on the eastern bank of the River Oder at Zehdenick, sixty kilometres north of Berlin. For three exhausting weeks the Brandenburgers managed to hold their positions against massed Russian attacks, despite the fact that battalions to the left and right of them had been overrun and destroyed. But, running low of ammunition, the 600 Battalion took a terrible battering and when they finally withdrew there were only thirty-six of the original eight hundred men still alive.

The survivors were reinforced by a few hundred trainees who had been rushed from Berlin in a desperate last effort to push back the Russian advance. Some wounded Brandenburgers rejoined their unit from their hospital beds, so strong was their commitment to their unit. Then the 600 Battalion was ordered to pull back to Neuruppin, some fifty kilometres to the west, and to defend the town to the last man. At dawn on April 3 1945 a single company of eighty-four Brandenburgers was facing two Russian tank divisions and two infantry divisions. It was an extraordinarily gallant defence by Special Forces soldiers under the most extreme battle conditions. The battle raged for eight hours as the Russians sent in wave after wave of tanks, backed up by hundreds of infantry.

After four hours the Germans had used up all their rocket-propelled weapons. In the final hours of this extraordinary battle the thirty surviving Brandenburgers had only hand grenades and satchel charges to hold back more than a hundred T34 and JS tanks. They still had some ammunition for their machine guns and personal weapons to keep the Russian infantry battalions at bay, but even that they had to fire sparingly.

When all their anti-tank rockets had been used, the Brandenburgers adopted a new tactic. They would wait in ditches until the first Russian tanks had passed by and some would then scramble out and on to the rear decks of the vehicles, dropping grenades into the open hatch to blast the crew. Others would run alongside the tanks, planting magnetic, hollow-charge grenades with a nine-second fuse on the sides. Having done this, the soldier would dive back into the ditch to escape the blast before moving on to the next tank.

Sometimes the Brandenburgers would wedge plate-shaped Teller mines between the tank's tracks and running wheels. These exploded with tremendous force, blowing the tank track apart and rendering the vehicle useless. Some soldiers stopped the advancing tanks instantly by simply flinging their satchel charges under the tracks. Within two hours more than sixty Russian tanks were at a standstill, wrecked by the audacious Brandenburgers.

Five separate assaults were launched by the Russian commanders and five times they were repulsed by the tiny band of men who were taking enormous risks, putting their lives on the line during every enemy assault. Because only some thirty Brandenburgers were alive after the fourth assault, they took up defensive positions only and used only machine guns, sub-machine guns and rifles. The Russian infantry had stopped trying to advance behind the protection of their tanks because they were being mowed down by the Brandenburgers. They let the tanks take the brunt of the German gunfire and waited for the inevitable victory.

Although they knew they were staring death in the face, the Brandenburgers held their ground and their nerve. Somehow they managed to stop the fifth tank assault, and the Russian crews leapt from their tanks and scrambled back to safety as bullets zipped around them. But such a one-sided battle could not last much longer. The sixth tank attack, at dusk, finally overran the German position and, ironically, it was at the very moment of defeat that the Brandenburgers' commander was given the order by radio to withdraw.

He had only about twenty men left. There were no wounded to take back. As the tiny band struggled away from the area in the darkness, they left behind the hulks of dozens of blazing or burnt-out Russian tanks. It was the final battle of World War Two for the German Special Forces. They had been utterly defeated, but in their defeat their remarkable courage could only be saluted.

CHAPTER 2

DESERT WAR

ONE OF THE MOST influential Special Forces units of modern times, largely because it spawned Britain's world-famous Special Air Service, or SAS, was the Long Range Desert Group, set up in North Africa during World War Two. Unlike the SAS, however, the LRDG was never intended to carry out offensive operations but to simply gather information about the enemy behind their lines. As the war intensified the LRDG did indeed carry out many offensive missions, harassing German forces in North Africa, but this was always a secondary role.

The needs of those who served in the LRDG provided a valuable guide to what the SAS would require to ensure that the new unit became a first-rate fighting force. In the case of the LRDG, those needs were basic but crucial, for most operations were carried out over weeks or months in desert conditions and in enemy territory, and self-sufficiency in everything from food and water to vehicles, fuel, weapons and ammunition was an absolute necessity. A further challenge was navigation, as there were no accurate maps of the deserts of North Africa and indeed scarcely any maps at all; the only navigational guides were the night stars and a compass.

One of the first offensive operations undertaken by the LRDG took place during the Eighth Army's major drive against the Germans in December 1941. The unit's officers were under orders to do whatever they could behind enemy lines to disrupt the Germans' efforts and distract them from their aims.

In his book *Providence Their Guide*, a history of the LRDG, Major General David Lloyd Owen writes:

One LRDG unit under Lt-Colonel Tony Hay came across a very inviting kind of target – a troops' road house – with some thirty enemy vehicles in the car park. That evening, as the sun went down, Tony led his patrol along a track which would join the main road about half a mile from the building. He drove fairly slowly towards it, and in doing so passed several vehicles going north along the road. They were mainly Italian but there was one anti-aircraft gun with a crew of four steel-helmeted Germans sitting to attention!

As they reached the building Tony Hay closed his trucks together and turned into the car park. At this moment they opened fire with their guns and hurled grenades into the trucks wasting no time in escaping from the hullabaloo that they had stirred up. It was getting dark, and he took his patrol off into the desert for the night.

Tony Hay and his small band of men in their eight vehicles continued to harass the enemy, setting fire to petrol tankers, knocking out aircraft on the ground, attacking convoys and terrorising the enemy by swift, night-time attacks on their camps and supply lines. The material damage caused to the Germans and Italians was not great, but these LRDG raids created an alarming effect on the enemy, who, more often than not, believed they were the advanced guard of a much greater British force heading towards them. Such operations gave the senior Axis officers many a headache. They were reluctant to order front-line soldiers to roam the desert searching for the British troublemakers, nor did Luftwaffe officers want their valuable aircraft to be seconded to searching the desert for pinpricks of nuisance.

Then, in 1942, the problem for the Axis desert armies intensified when the SAS and the LRDG joined forces to create a brilliant partnership. The plan was that the LRDG, the more experienced partner, would take several SAS units of four men to carry out raids, sabotage and sometimes parachute operations against the Axis forces. The main aim of both Auchinleck, Commander-in-Chief Middle East Forces, and his successor Montgomery was to knock out as many German aircraft as possible because these were having a devastating effect on Allied equipment and troops.

One of the first joint battle plans was to attack two German airfields at Sirte and Tamet, some three hundred and fifty miles from Jalo in Libya. Major Paddy Mayne, a former Irish rugby international, was in command of one patrol of four vehicles and eight men which made its way under cover of darkness to within three miles of the target at Tamet.

Major General Lloyd Owen again:

Mayne found himself with his men, all carrying some 70lbs of explosives on their backs, in full view of an airfield with aircraft parked all around the edges... Paddy Mayne was to take full advantage of the setting.

At the western edge of the airfield Mayne had seen some huts where the aircrews were living, and after dark he decided to deal with the occupants of these before turning his attention to the aircraft. Paddy waited until he thought most of them would be asleep; then he and five others rushed into the huts and with bursts from tommy-guns fired from the hips they made quite sure that there would be no one left alive to prevent them dealing with the aircraft.

He wasted no time. The whole raid only took about quarter of an hour, but this was the SAS method of working, and when they were on their way to the rendezvous a total of twenty-four aircraft and the fuel dump were either blazing or ready to explode into flames.

In fact the powerfully built Mayne personally destroyed one

aircraft with his bare hands, climbing into the bomber's cockpit and tearing out the instrument, which he took back to base as a souvenir. In those raids on that single night sixty-one enemy aircraft were destroyed on the ground. Effusive praise for the daring and brilliance of the joint SAS and LRDG operations came the following day. The fame of the SAS, which has never yet diminished, was born.

The idea of the SAS had come to its celebrated founder, Colonel David Stirling, as he lay, injured in a parachuting accident, in a hospital in Alexandria in the summer of 1941. In a memo written in pencil on reams of paper he argued for deep-penetration operations behind enemy lines in which small units would carry out strategic raids after parachuting into enemy territory. He believed that a small force of saboteurs could inflict a level of damage on enemy airfields equivalent to that of a Commando squad twenty times greater.

Stirling was a Cambridge-educated Scots Guardsman who joined one of the first army Commando units in 1940. Six foot five inches tall and weighing sixteen stone, he had always sought adventure and his principal hobby was rock climbing and mountaineering. The outbreak of war in 1939 gave Stirling the chance of as much adventure as any man could ever want.

Now, as a lowly young subaltern, he faced the formidable task of trying to sell his revolutionary idea to his illustrious army commanders, the only men who had the power to put such ideas into operation. It would not be easy. Knowing it would be all but impossible to gain an interview with the Commander–in–Chief, he decided on a frontal attack. Though still on crutches, Stirling decided to 'break in' to the C-in-C's headquarters, situated in GHQ Cairo, and beard Sir Claude Auchinleck in his lair. He threw his crutches over the perimeter fence and then, somehow, clambered over, only to set alarm bells ringing.

As the Military Police searched for the intruder, Stirling reached the Commander-in-Chief's block and hobbled into the office of his deputy, Lieutenant General Neil Ritchie, just as the MPs were about to grab him.

'I think you had better read this, sir,' Stirling said, handing the surprised Ritchie his memo. Still on his crutches, Stirling then withdrew, helped by the MPs. To his amazement, three days later he was recalled to talk over the idea with both Ritchie and Auchinleck, and within weeks he was promoted to the rank of captain and told to recruit and train sixty-six Commandos for his revolutionary idea. The SAS Regiment was born.

Later Stirling would write about the guiding principles of his brainchild:

Strategic operations demand, for the achievement of success, a total exploitation of surprise and of guile. A bedrock principle of this new regiment was its organisation into modules or sub-units of four men. Hitherto, battalion strength formations, whether Airborne formations or Commandos, had no basic sub-unit smaller than a section or a troop consisting of an NCO plus eight or ten men... In the SAS each of the four men was trained to a high general level of proficiency in the whole range of the SAS capability and, additionally, each man was trained to have at least one special expertise according to his aptitude. In carrying out an operation – even in pitch dark – each SAS man in each module was exercising his own individual perception and judgement at full stretch.

Stirling developed the idea of a module of four men, or 'brick', in order to prevent the emergence of a leader of operations. An important result has been to foster a military democracy within the SAS, which, through ruthless selection of volunteers, has traditionally led SAS soldiers to motivate and discipline themselves. Stirling's revolutionary idea of the four-man brick proved so successful that it has been copied by nearly every other Special Forces unit in the world.

He described the SAS Regiment's philosophy thus:

1. The unrelenting pursuit of excellence.

2. The maintaining of the highest standards of discipline in all aspects of the daily life of the SAS soldier.

3. The SAS brooks no sense of class and, particularly, not among the wives. This might sound a bit portentous but it epitomises the SAS philosophy.

The SAS's first parachute operation, during a sandstorm in the desert, was a complete disaster in which both aircraft and parachutists were swept away. But the second operation, undertaken in collaboration with the Long Range Desert Group, was a brilliant success, destroying more than one hundred Luftwaffe aircraft on the ground at a number of German airfields. And what was remarkable about this mission was that it had been conducted by only twenty SAS men. Auchinleck was very happy that he had given his support to the young upstart Stirling and his radical idea. More importantly, that first series of attacks had established beyond doubt the role of the SAS. This new Special Force would spawn, throughout the world, many other similar units which have proved their worth ever since.

The extraordinary success of the young SAS came to the attention of Hitler, who dispatched a personal order to General Rommel: 'These men are dangerous. They must be hunted down and destroyed at all costs.'

And they were. In his envy, anger and total disregard for the Geneva Convention's pronouncements on prisoners of war, Hitler decreed in the early months of 1945, when defeat for Germany had become a reality, that every SAS man captured and imprisoned should be tortured and then killed. Indeed his brutality went further. A team of British investigators discovered that, after suffering torture, some SAS men had been flayed to death, others had been roasted alive on a spit.

The SAS was disbanded at the end of World War Two, but

those soldiers who remained in the unit made it plain to the military authorities that they had joined it for action and adventure, not to sit on their backsides at home twiddling their thumbs. Given the SAS's dramatic intervention in many operations during the war, the Ministry of Defence came to recognise that such men could be thrown into the deep end in virtually any military situation. As a result, the SAS began to be used in countless operations where small groups of men could carry out the task with great speed and commitment and, importantly, without the need for supply lines and logistical support.

In fact the services of the SAS were required within a few years of the end of the war. In Malaya, the newly armed communist guerrillas were indulging in a riot of murder, and the British government had to act to protect both the British families who had made the country their home and the indigenous population. By March 1950 the terrorists had killed eight hundred civilians, three hundred police officers and a hundred and fifty soldiers. Something had to be done.

A Special Force along the lines of the disbanded SAS was put together from soldiers and reservists who volunteered, but this was not particularly successful until discipline was tightened and the new SAS groups began to live for periods of weeks, or sometimes months. in the jungles of Malaya. For some six years the SAS maintained a presence in the country, taking part in all the most dangerous missions, tracking terrorists for days or even weeks. By 1956 five SAS squadrons, with a total of five hundred men, were playing an important role in the conflict, and the numbers of deaths in the jungle at the hands of the terrorists had been cut to about one a week. But there were still some two thousand hardened guerrillas hiding out in the jungle and swamps and terrorising villagers and farmers.

By the end of the 1950s the insurgents had been teasing the government forces, and killing or threatening the indigenous population, for almost ten years. Employing great technical skill, the SAS searched them out and destroyed them and their base camps, striking characteristically hard and fast. Now they were finally gaining control, and by 1959 the terrorists realised that they were not going to win the war in Malaya. Slowly but surely, the SAS were forcing them into retreat. The game was up, and groups of terrorists began to surrender. A war which many believed could not be won against such a guerrilla force had indeed been won by the SAS. Military strategists believed that it was the ability of the SAS to alter their tactics and adapt a new approach to jungle warfare that had brought them victory.

But it was not the end. The war in Malaya took on a totally different and more serious aspect when Indonesia decided to intervene in a bid to destabilise the fledgling Malaysian Federation by attacking Brunei, Sabah and Sarawak, to the north of Indonesian Borneo. The Indonesians had a large, professional army trained in jungle warfare and began to use members of this for clandestine guerrilla operations against Malaysian forces in Borneo. Almost unknown to the rest of the world at the time, a war lasting four years was fought by some thirty thousand British, Australian, New Zealand and Gurkha troops against an army of more than a hundred and twenty thousand Indonesians. And the SAS were in the thick of it.

Their task was deep penetration and intelligence gathering in the jungles of Borneo behind Indonesian lines. SAS bricks would be sent to watch and wait, and then to report back and advise those officers at headquarters whose job it was ambush the Indonesian forces or hit them in fast counter-attacks from the air and roads.

After commanding the SAS in Borneo, General Sir Walter Walker commented:

'I believe that a few SAS bricks were equal to a thousand infantry not because they had the equivalent fire power but because their Intelligence gathering could save that number of lives in battles won without a fight.'

It was an extraordinarily hard life for the SAS men. They would have to live rough for three weeks at a time, surviving on poor, hard rations and very little sleep, and maintaining almost complete silence. They were taught to conceal themselves in the jungle by never smoking or chewing gum, never washing with soap, never cleaning their teeth, and never sneezing or coughing. When they returned to base they were often dehydrated and had usually lost twenty to twenty-five pounds in weight, so that they looked like skeletons. They were permitted just five days' recuperation before their next three-week sojourn in the jungle. The three arduous years of this conflict, from 1963 to 1966, gave SAS volunteers an invaluable experience of jungle warfare.

Tasked with searching out, hunting down and killing Indonesian troops who had infiltrated into Malaysia along a seven hundred-mile border of jungle, the SAS knew that the only way to police this border was to win the hearts and minds of the Sarawak villagers, the way they had done in Malaya. As a result, the villagers became the eyes and ears of the SAS bricks, feeding them information about the enemy. Without such help it would have been impossible for the SAS and their Gurkha comrades to carry out the allotted task. SAS men would move cautiously into a village bearing gifts of medicine, small portable radios and other nick-nacks and make friends with the families, sometimes staying for six months at a time and becoming useful members of the village. In return, the tribesmen provided valuable information which was relayed back to base. The war came to an end in 1966 after President Sukarno of Indonesia was sidelined by his generals. British casualties throughout the entire period were nineteen killed and forty-four wounded; forty Gurkhas were killed and eighty-three wounded. The Indonesian death toll has been conservatively estimated at two thousand.

In the late 1950s another area of conflict had opened up – in the Middle East and, in particular, southern Arabia – and the SAS would operate there until the 1970s. Early on they were sent into Aden and Oman for a series of hellish operations often referred to as the last wild colonial war. This author served in the private army of the Sultan of Muscat and Oman. Given one hundred and twenty Arab and Baluchi troops, three three-tonners, a jeep and three excellent camels, I had to patrol the desert west of Muscat City and intercept caravans of camels transporting rifles and ammunition to the rebels fighting the Sultan in the famous Empty Quarter, the mountainous region on either side of the Saudi Arabian border. We would ambush rebels and their camel trains, raid their camps in the hills and stop and search anyone we came across in the desert. Every few days we would be attacked by marauding rebels – usually at night – and there would be a firefight lasting perhaps twenty minutes. Never once did we have to pull out of a battle but would usually see the rebels off by sending a five-man patrol to one flank or the other to surprise them. It worked every time. And in the six-month period of operations in the desert we suffered only two men wounded, and they were only flesh wounds which I treated and bandaged.

It was nothing like the tough, hard life the SAS had to endure in the region against substantial and formidable guerrilla forces. In December 1958, in one of the toughest assignments ever given to the unit, two troops of SAS men began the task of rooting out and destroying the main guerrilla headquarters and weapons and ammunition store in the caves of the Jabal Akhdar, the Green Mountain. From this remote, eight thousand-foot hideout in northern Oman, guerrillas had been leading marauding parties against Omani villages and camel trains in an attempt to destabilise the Sultan's autocratic rule and set up a communist republic.

As the eight SAS men made their way under cover of darkness towards the summit, climbing up a near-vertical rock face, gunfire rained down on them. They were under attack from some forty guerrillas armed with Bren guns and rifles. The onslaught was ferocious and there was virtually no cover, so the SAS men lay still, held their fire and waited. Confident they had killed the raiders, the guerrillas made their way down to where they had seen movement. Only when the enemy were about one hundred and twenty yards away and visible in the star-studded night sky, did the SAS open up with automatic fire. Five guerrillas were killed outright and four wounded. The rest fled for their lives.

Two weeks later, while still trying to penetrate the mountain headquarters, Captain Rory Walker was leading another exploratory raid near the top when his two troops of SAS men were once again detected just before they reached their goal.

An SAS soldier was using a rope to climb up a fault in the cliff when a guerrilla above him shouted, 'Come on, Johnny' and opened fire. Walker took the man's place, climbed up the rope and hurled a grenade over the top into the group of guerrillas gathered there, killing one and wounding some of the others. Walker signalled to the rest of his men and continued up the rope. When they had all reached the top they charged the enemy, firing their Bren guns from the hip as they ran. Eight more guerrillas were killed in the firefight; the rest fled into the caves.

On the other side of the mountain, also controlled by the enemy, Captain Peter de la Billiere, with two SAS troops, made a ten-hour forced march through enemy-held territory in darkness. In his book *Who Dares Wins*, Tony Geraghty writes:

The SAS crept to a point two hundred yards from the cave mouth (where weapons and ammunition were stored) lined up a 3.5 inch rocket-launcher, and waited. The only point from which the SAS could open fire was below the cave, and this meant that the rocket crew had

to kneel or stand to use the weapon. The same firing-point was, for want of something better, a natural amphitheatre whose upper slopes were honeycombed by many small caves sheltering enemy snipers. At dawn, as the first of the guerrillas emerged, stretching his sleep-laden limbs and yawning, the soldiers poured a hail of missiles and machine-gun fire into the main cave.

Describing the action, one SAS officer later wrote:

Even such withering fire did not cause the rebels to panic or surrender. They quickly dropped into fire positions and returned the best they could. Reloading and firing the 3.5 inch from the standing position became interesting. What made it particularly interesting, as well as infuriating, was the failure of many of the SAS to leave the launcher after being fired. They remained unfired but 'active' and had to be extracted immediately and replaced with another round, regardless of the necessary safety drills.

The noise of the battle instantly brought down rifle fire from the surrounding hills. Outlying rebel pickets retreated slowly and the SAS picked them off one by one. The rebels still had a mortar firing from a crevice behind the cave, but the SAS laid on air support. As RAF Venom aircraft came swooping in, one of their rockets made a direct hit. Mortar and men were destroyed immediately.

This action now became a fighting retreat, in which men moved back singly or in pairs, using every scrap of cover available. This lasted rather more than fifteen minutes and was covered by a .3 inch Browning machine-gun manned by a regimental veteran.

But the battle on the Jabal Akhdar was not yet won. In addition to an SAS squadron, the conflict now involved an infantry troop of Life Guards, two troops of Trucial Oman Scouts and a few signallers and REME men – a total of some two hundred soldiers. And the conditions were harsh. The winds were incessant and bitterly cold by day and night, when even the water bottles froze during the night.

By the end of December, when the SAS had been on the

mountain some two weeks, the commanders knew there was no way they could dislodge the guerrillas, who were in a strong defensive position. One very good reason for the confidence of the defenders was the fact that, historically, the Jabal Akhdar had never been conquered by any force. It was considered to be impregnable. The SAS needed more men if victory was to be achieved.

The newcomers, a further SAS squadron of one hundred and twenty men, were given the task of staging a frontal attack between the twin peaks of the mountain. The assault route was up a narrow track that climbed four hundred feet to a fortified and very easily defended position. However, this assault was a feint, and the main attack would come from the other side of the mountain. The SAS men had to climb for some nine hours in the darkness up a four thousand-foot slope, for which ropes were needed on one traverse. Never before had any force attempted to scale and conquer the mountain from this angle of attack. Each soldier carried on his back sixty pounds of equipment, mainly ammunition. Behind came donkeys, laden with heavy machine guns.

With dawn only an hour away, the SAS men found themselves still some distance from the mountain top. They now faced two choices. They could either keep moving slowly and steadily to the top and pray they wouldn't be spotted by the guerrillas on guard, or they could dump their heavy ammunition packs and make a dash for the top in the hope of catching the defenders unawares. The second option was a major risk. If it went wrong they could all be wiped out in a blaze of automatic fire from above. And yet they decided to take that risk. This was an attack on a well-defended position which tacticians believed was impregnable. One of the most daring and outrageous assaults ever undertaken by a small group of men, it was to make the SAS into the legendary fighting force the world has come to admire and respect.

They dumped the ammunition and began the suicidal scramble up the final three hundred metres of the steep mountain. There was no cover whatsoever – no trees, no shrubbery, no gorse, nothing to prevent a lookout seeing these forty men with their blackened faces as they struggled, at times on all fours, up the mountain. Ten minutes later they reached the plateau at the top, exhausted and so breathless they couldn't speak. And there was not person in sight.

Fearing a trap, the SAS men took cover, waited for the anticipated onslaught and called up RAF Venom strike planes to bomb the caves a few hundred metres ahead of them. The Venoms responded and also dropped more ammunition. It seemed that the guerrillas had feared a full-scale airborne invasion and fled their stronghold, leaving behind mortars, heavy machine guns, Bren guns, mines and ammunition. It was indeed a famous victory for the SAS.

The Oman campaign, at first little more than a daredevil series of skirmishes, would, however, point the way towards a new type of war for Britain as it divulged itself of bits of the empire in various parts of the world. From now on British government military policy would focus on strategic mobility, with the SAS and other arms of the British forces acting more like fire brigades racing to douse fires wherever they might break out.

In 1972 a small group of SAS men fought a remarkable old-fashioned battle in the Omani coastal village of Mirbat, a place with two small, crumbling old forts, some fifty flat-topped houses and surrounded by barbed wire. However, Mirbat was only a hundred miles or so from South Yemen, then a haven of Arab communist guerrillas armed with AK47s and other Soviet weaponry, who wanted to overthrow the Sultan of Oman and take control of the country, which they suspected was rich in oil. Small groups of SAS units were in place to support the Sultan's own forces.

The battle was part of a secret war being conducted by British SAS men in Oman. This war began in the late 1960s and lasted into the middle of the following decade, but at the time the British public had no idea it was being waged by their government. In fact it was a war of which Britain should have been proud and in which SAS soldiers showed great courage and skill fighting far larger forces of determined Soviet-backed rebels. This is the story of one of those awesome battles.

A group of ten SAS men and thirty Omani soldiers, the latter armed with old-fashioned British Lee Enfield .303 rifles, were protecting the villagers of Mirbat and their livestock against occasional mortar bomb attacks from the unseen enemy on the other side of a small range of hills. The SAS had their personal weapons plus a World War Two twenty-five-pound gun, a single Browning machine gun and a single eighty-one-inch mortar. However, in July 1972, the Yemeni guerrillas decided for the first time ever to launch a major attack against Mirbat in a direct frontal assault. Two hundred and fifty guerrillas, armed with anti-tank rifles, rocket-launchers, mortars, heavy machine guns and AK47s, launched a silent attack early one morning before first light. They surrounded the village while everyone asleep and then opened fire, taking both the villagers and the SAS men totally by surprise.

The battle began when dozens of mortars rained down on the village, waking the British soldiers and everyone else. Nothing like this number of mortar shells had ever been fired before. Within minutes the SAS men were in action – two firing the mortar and two firing the Browning machine gun, while the other six kept up precision firing with FN rifles and the light machine gun. The Omani riflemen also took up positions around the perimeter of the village, maintaining a steady stream of fire. A team of three, including two Fijians, manned the twenty-five-pounder, raining shells on the advancing guerrillas. For the next hour the defenders fought a frantic battle, pouring as much automatic fire as possible at the guerrillas as they advanced relentlessly towards the village. The guerrillas, who appeared to be well led, were likewise directing as much firepower as they could at the village.

As the battle raged the women and children trembled in their houses. The noise of gunfire was almost deafening, the explosions of the rockets as they hit their targets loud and unnerving. Never before had the SAS men been subjected to such an onslaught from such a numerically superior force, but their discipline and their reactions never wavered. Their exceptionally tough training showed through.

Occasionally there were lulls in the firing, only for the rebels to launch further withering attacks some ten minutes later. Within three hours the Yemeni guerrillas were within ten yards of the perimeter wire, at three separate places, and showing extraordinary courage in the face of sustained automatic fire from the defenders. From two positions close to the village the guerrillas opened fire on the old fort with rocket-launchers and Soviet RPG rockets. Things were looking desperate for the SAS. There was no chance of air support, partly because the cloud level was low, but also because the attackers and defenders were too close to one another, almost locked together with only a few yards between them.

Nevertheless, the RAF at Salalah, the provincial capital, sent in a helicopter with the intention of taking out the wounded. But the chopper could not land on the small site near the sea and, as bullets pock-marked it, was forced to fly off again.

It was at this point that Captain Mike Kealy, who was in command of the Mirbat operation, realised that the heavy gun and the Browning machine gun had ceased firing over at the old fort. The twenty-three-year-old officer knew that unless he could get these weapons firing once again the entire place would be overrun in no time and the battle lost. He and his medical orderly, Trooper Tom Tobin, ran the three hundred yards from their position in the headquarters building to the Old Fort, stopping every few paces to fire their automatic rifles at the enemy at the wire before sprinting on again, while the rebels poured firepower all around them. Unbelievably, neither man was hit.

The situation they found in the fort was desperate. The Omani soldiers had put up a fantastic fight, but four were lying dead; others were bleeding badly from bullet wounds. One had had half his face blown away; another had lost an arm; another had a gaping hole in his stomach; and yet another's legs were missing. Yet somehow even the wounded men kept firing at the enemy. Within a minute of Kealy's arrival at the gun emplacement the rebels were close enough to begin lobbing grenades. He knew that if one grenade landed in the gun pit they would all be dead. One did land, but it failed to go off.

Before making his dash to the gun emplacements, Kealy had radioed Salalah to advise them that he had no idea how long they could hold out unless RAF Venoms could launch a strike against the enemy. He warned of the dangers to his own men and the villagers, but stressed that there was simply no alternative.

The close-quarter battle around the Old Fort and the bunkers had been raging for some two hours when, out of the cloud, came two RAF Venoms, flying at just one hundred feet. By relaying radio messages to the pilots, Kealy directed the bombs to a ditch where many of the rebels were gathered, and to a machine-gun emplacement only about fifty yards from the village which was pouring in non-stop heavy fire, causing many casualties. The Venoms' timely arrival seemed to terrify the rebels, many of whom turned and fled.

But the battle of Mirbat wasn't over yet.

It was the pilot of one of the Venoms, which was badly hit by machine-gun fire, who reinforced Kealy's earlier pleas for help. 'My God,' he said over his intercom, 'there are hundreds of them down there!' That single comment made the commanders in Salalah realise that unless major reinforcements were rushed to Mirbat there was very little chance that the SAS men could hold out for much longer. It was clear that the base was about to be overwhelmed.

Then an amazing stroke of luck came to Kealy's rescue. The previous day G Squadron SAS had arrived at Salalah ready to take over from Kealy's C Squadron and were on the firing range at Salalah testing their weapons when the urgent requests for assistance came over the radio. Within an hour eighteen men of G Squadron had packed into two helicopters and flown at sea level to Mirbat.

The men hit the ground running and, after a quick briefing from Kealy, advanced out of the fortified village in pursuit of rebels sheltering behind a ridge some three hundred yards away. After firing a few rounds the rebels turned and fled, with the fresh, super-fit SAS men on their tails. Two more helicopter flights brought in more SAS men, who advanced on the beach, where rebels had been keeping up a steady stream of fire throughout the battle from three defensive positions.

These rebels, numbering about one hundred, had to be removed, and quickly, to bring the situation under control and restore some sort of calm among the beleaguered villagers. The guerrillas fired their machine guns and lobbed grenades at the SAS, but after ten minutes of a fierce firefight the SAS had moved into positions which forced the rebels to either surrender or risk being shot to pieces. They surrendered.

The battle had claimed the lives of two SAS men, and two were seriously injured. Two Omanis were killed and a further two seriously wounded. The rebels left thirty bodies behind and ten more were taken prisoner. Some twenty more guerrillas had been severely wounded. When the war in Oman ended, some four years later, a tribute was paid to those SAS men who had fought at Mirbat. A guerrilla commander said that they had never recovered from that defeat and were never able to stage any similar attacks again.

Within the emerging military world of Special Forces the courageous exploits of the SAS in Oman were seen as a measure of their competence, skill and determination in the face of overwhelming odds. And there would be more battles which only came to the notice of the general public some years later. The SAS were proud of the fact that the regiment was almost a secret service, their exploits known to only the top military brass, the Secret Intelligence Services and officials at the Ministry of Defence.

The closely guarded world of the SAS would be blown wide open in May 1980 when television news cameras showed live coverage of men in black clothing and balaclavas wielding sub-machine guns and machine pistols on the roof and window sills of the Iranian Embassy in Prince's Gate, London, a stone's throw from the fashionable shops and restaurants of Knightsbridge.

Most TV viewers watched in silence, spellbound by the scene unfolding before them. They saw these men place objects against windows of the embassy, and within seconds two explosions filled the air with noise and dust. Seconds later viewers watched as more men in black fatigues abseiled down the rear of the building. More explosions followed as stun grenades were thrown into the rooms, and from inside the building came the muffled noise of machine-gun fire.

The nation was stunned. Who were these men dressed all in black invading an embassy in the heart of London? People knew that some gunmen had been holding officials of the Iranian Embassy for several days. But to most British viewers this action by unknown men was completely alien. Within hours, however, all would become clear, and over the following days television and news media would explain in great detail that these men were members of Britain's SAS, a secret elite force responsible for counter-terrorism. The existence of the SAS had been known about since the days of World War Two, but this assault brought home just what skill they possessed. Overnight the TV images were flashed around the world and from that moment on the SAS lost its secrecy and gained, instead, fame and notoriety as the world's most daring, courageous and efficient fighting force. Nearly every SAS man before and since that fateful day would have preferred the unit to have remained behind the scenes, out of view and out of the headlines. It was not to be.

The events of May 5 had begun a week earlier, when six young Arabs from Khuzestan province in Iran walked towards the Iranian Embassy carrying machine pistols, 9mm pistols and Russian fragmentation grenades beneath their coats. Their mission was to bring the struggle for the independence of greater Khuzestan – to be renamed Arabistan – to world attention and to humiliate the Iranian government.

On duty guarding the embassy that day was PC Trevor Lock, armed with a .38 revolver. He opened the front door to be confronted by the six men, who were all wearing Arab headscarves over their heads and faces. Slamming shut the door, PC Lock managed to sound the alarm on his personal radio as the gunmen opened fire, shattering the glass door and blinding him with flying splinters of glass. Firing their weapons, the gunmen rampaged through the building, rounding up everyone and locking all twenty-six hostages in Room 9 on the second floor.

By the end of the first day communication with the terrorists had been established by a secure landline and the area cordoned off by armed police. The BBC had also been requested by one of the hostages, directed by one of the terrorists, to make known that the group were demanding the release from Iranian jails of ninety members of the Democratic Revolutionary Movement for the Liberation of Arabistan. Three hours later the BBC received another call from the terrorists, this time with a threat: 'Unless the Iranians free the DRMLA prisoners by noon the following day the embassy will be blown up and the hostages killed.'

In his book *Secret Soldiers* Peter Harclerode details the dramatic sequence of events in the SAS's involvement, which began just hours after the hostages were seized. Before dawn the following day Lieutenant Colonel Michael Rose, Commanding Officer of 22 SAS, was in a building next door to the embassy, surveying the scene and planning an operation for immediate action if the terrorists did start killing the hostages. The SAS planned to make an entrance through the roof of the embassy and upper floors and then clear the building of terrorists from the top downwards, in the hope that they could reach the hostages before they were massacred.

Two days later the terrorists inside the embassy were becoming agitated as none of their demands had been met, and the gunmen informed PC Lock that, as a result, they would shortly have to begin shooting the hostages. In fact they appeared to be wavering in their resolve, for they now demanded that the group be allowed to leave Britain for an unspecified destination in the Middle East with only a few of the hostages. The remaining hostages would be left in the bus when the group boarded the aircraft at Heathrow.

Unbeknown to the terrorists, however, Prime Minister Margaret Thatcher and her senior advisers had summarily dismissed their demands for safe passage. On the fourth day of the siege one of the demands was carried out – the BBC broadcast a statement dictated by the group's leader, who went under the name Salim, a highly intelligent twenty-seven-year-old member of a middle-class Khuzestan family, university-educated and fluent in English, German, Arabic and Farsi. Two hostages were released and, in return, the police sent in a meal from a Persian restaurant. The terrorists and the hostages were jubilant, believing there would now be a peaceful outcome to the raid.

But they did not hear two SAS men on the embassy roof, who silently removed the glass from a skylight, opened the skylight and then carefully put everything back in place, to provide access for a possible assault.

When nothing happened the following day, however, the terrorists once again became jittery and fractious, convinced that police or members of the armed forces had gained access to the building. The hostages also became despondent and fearful of their future. Worried that the terrorists' behaviour was becoming increasingly irrational and that they appeared more desperate than ever, PC Lock persuaded Salim to let him talk to the police negotiators. PC Lock left the negotiators in no doubt that the terrorists' patience was exhausted and time was running out.

An hour later Salim told the police on the secure phone link: 'Bring one of the Middle East ambassadors here to talk to me in forty-five minutes or I will shoot a hostage.'

The terrorists reinforced this message by bringing PC Lock to the phone and confirming that a hostage had been selected and would be shot unless Salim's demand was met. The minutes ticked by and no response came back from the police, who were in fact taking their instructions from COBRA – the Cabinet Office Briefing Room, a crisis committee set up to oversee any major emergencies within the UK and chaired by the Home Secretary and representatives of the Foreign Office, the Home Office, Ministry of Defence and the security and intelligence services.

Harclerode wrote:

Inside the embassy meanwhile Abbas Lavasani, one of the embassy's two press officers, had been taken to the ground floor and,

his hands bound behind his back, tied to the banisters at the foot of the stairs. At 1.45 pm Salim telephoned the negotiators and held the receiver next to Lavasani who identified himself as one of the hostages. No sooner had the negotiators heard him say his name than they heard another voice cut in, shouting, 'No names! No names!' Immediately afterwards came the sound of two or three shots followed by a long choking groan. Salim then came on the line again and announced that he had killed a hostage.

This led to feverish activity, for the killing meant that COBRA had misread the situation. Furthermore, now that the terrorists had resorted to killing hostages, experienced hostage negotiators warned that there was every probability that more shootings would follow. As COBRA's top officials and military advisers decided what action to take, the terrorists announced that another hostage would be shot in thirty minutes' time.

In a desperate effort to calm the terrorists, a senior officer of the Anti-Terrorist Squad persuaded Dr Sayyed Darsh, the senior imam at the Central Mosque in Regent's Park, London, to act as a mediator in an effort to save the lives of the hostages. At the same time a meeting of all Arab ambassadors in London took place, and they agreed to issue a press release saying they would also act as mediators. But it was all too late.

No Arab ambassador was ready to hold talks with the terrorists and the attempt by Dr Darsh to calm Salim by reading from the Koran seemed to infuriate rather than placate the terrorist leader. Over the phone the sound of three shots rang out. A few seconds later the front door of the embassy opened and a body was dumped unceremoniously outside.

News of the second shooting was flashed to COBRA, where the Home Secretary, William Whitelaw, telephoned Mrs Thatcher requesting permission for the SAS to carry out an immediate assault. At 7 pm control over the assault phase of the operation was passed to Colonel Rose.

Harclerode again:

At 7 pm as Salim was still talking to negotiators on the phone two large explosions reverberated through the embassy building and the assault began.

The plan called for all floors of the embassy to be attacked simultaneously by the two SAS teams, Red and Blue. Red was responsible for clearing the top half of the building with two groups of four men abseiling in two waves from the roof to the second-floor balcony at the rear of the building where they would force an entry via three windows. Meanwhile, another group would assault the third floor using a ladder to descend from the top of the building to a lower roof. The top floor would be cleared by another group entering via the skylight.

Blue Team was tasked with clearing the lower half of the building comprising the basement, ground and first floors. Explosives would be used to blast through the bullet-resistant glass of the French windows at the rear of the building and the windows on the first-floor balcony at the front. Members of Blue Team would also be responsible for firing CS gas canisters through the second-floor windows...

Dressed in black assault suits and wearing body armour and respirators, each man in both teams was armed with H&K MP5 sub-machine guns and a Browning 9mm pistol. In addition, one group in each team was equipped with a frame charge manufactured from linear cutting charge explosive mounted on a light wooden frame. As the teams took up their positions, an explosive charge was lowered and suspended just above a glass roof covering the building's stairwell...

The charge above the stairwell's glass roof detonated with a huge explosion, blowing it in and sending a shock wave through the building. At the same time, members of Blue Team began firing CS gas canisters through the second-floor windows while the rear assault group smashed their way through the French windows with sledge hammers.

Television viewers saw members of Blue Team clambering

across from the adjacent house on to the embassy's first-floor balcony. One SAS man put the charge against the window, blowing it away, and then threw a stun grenade into the room. The four SAS men clambered inside, guns at the ready. They knew that it would become a case of 'them or us' and that their reactions would be decisive.

Two SAS men moved from the room on to the landing, but there was no one there. They heard noises of a scuffle coming from a side room and burst in to find PC Lock struggling with Salim. On hearing the explosions, the policeman had gone for the gun he had concealed throughout the six-day siege, but Salim had seen him fumbling for the gun and was desperately trying to rip the handgun from his grasp. One SAS man yelled at PC Lock to 'get out of it' as both soldiers opened fire with their MP5s, hitting Salim in the head and chest. He died instantly.

The other two SAS men who had gone in through the front windows opened the door to the smoke-filled Ambassador's office to be confronted by an armed terrorist. A burst of automatic fire sent the gunman staggering backwards, and he disappeared into the smoke. Cautiously, the two SAS men moved into the room, unable to see more than a few feet. They found the gunman lying on the sofa with his weapon at the ready. He died in a hail of bullets from their sub-machine guns.

The SAS troopers had not yet discovered the exact whereabouts of the hostages, but suspected they were being guarded by the remaining three armed terrorists. They knew they had only seconds to save their lives. One of the SAS men at the back of the building looked through a window to see a gunman lighting newspapers in a crowded room. He smashed the window, threw in a stun grenade and clambered in. The terrorist ran from the room and across the landing to Room 10, where the hostages had been herded. As he entered he opened fire on the huddle of frightened people with his machine pistol and was joined by another terrorist, who also opened fire indiscriminately at the screaming hostages.

A moment later the SAS man ran into the room and shot one terrorist in the head with a single bullet from his Browning. Ali-Akbar Samadzadeh, one of the embassy's press officers, had been shot dead and Dr Ali Afrouz had been shot twice in the legs. There was chaos in the room, with the SAS yelling at the hostages to lie down; the hostages screaming in fear; and the terrorists shouting for mercy as they threw themselves down on the floor among the hostages. The SAS men had a problem because there was no way they could tell – in a split second – the difference between the Arabistan terrorists and the Iranian hostages. The SAS men shouted at the hostages to get out of the room as they searched out the terrorists, fearful they might make one last desperate effort to kill them. One man lying on the floor made a sudden move and was shot in the back. Under his body they discovered a grenade.

The soldiers were still desperately trying to identify the remaining two gunmen. They lined up on the landing and hurriedly and roughly pushed the hostages down the stairs while checking each of them. One man was being pushed towards the top of the stairs when a trooper recognised his face. The SAS man glanced down and saw that the terrorist was holding a grenade and he shoved him in the back as hard as he could so that he stumbled down the stairs. He had not wanted to open fire because the stairs were crowded with hostages. Another SAS man standing halfway down the stairs clubbed the terrorist on the back of the head as he toppled past him, and the man landed at the bottom of the stairs in a crumpled heap. A second later the SAS soldier guarding the hall shot the man in the head and took away the grenade. The safety pin was still intact.

But the SAS squad were certain there was still one remaining terrorist unaccounted for. They ordered everyone into the rear garden and made them lie face down on the lawn while Blue Team soldiers handcuffed each of them. Within a couple of minutes the hostages identified the remaining terrorist and he was pulled to his feet, searched and handed over to police. The hostages were released but ordered to remain in the garden while the entire building was searched and declared safe by the soldiers. Only then did the SAS men leave, driving away in unmarked vans back to obscurity.

Those sensational TV pictures, flashed around the world, made the SAS's reputation. From that day the unit had become the world's foremost anti-terrorist organisation, praised for its courage and professionalism. The action had sent a clear signal to all terrorists that Britain would deal firmly with any threats to the community. It also brought many requests from governments around the world seeking the SAS's assistance in training their own anti-terrorist squads. From the viewpoint of the SAS, however, the publicity was most unwelcome. The last thing the secret force needed was to be the centre of attention amid a blaze of publicity.

CHAPTER 3

'THE LEGION IS OUR COUNTRY'

SINCE ITS INCEPTION in the 1830s, the French Foreign Legion has been involved in many of the same kinds of military operations as today's Special Forces. However, unlike nearly every other such unit, the Legion is a mercenary force. It is defined in this way because France is not the country of birth or adoption of some of its members, and these soldiers serve only for the pay they receive. However, since the pay has never been high, men join the Legion less for the money than for a wide variety of other reasons.

Before anyone signs up as a Legionnaire he knows that the discipline will be severe, the training pitiless and that his comrades may well be men who are on the run from justice in their own country. For many years the Legion had the reputation of being not only tough and hard but also a unit where the majority of men had enlisted for reasons other than wanting to serve in an elite military force. In the past some volunteers were described as murderers, brutes or simply hard bastards. And throughout the Legion's history some have joined up because they wanted to make a fresh start in life, whether for personal or emotional reasons, while others have done so to prove to themselves they are tough.

But after the war in Algeria, which ended in 1963, the French government drew up new rules rejecting volunteers who had a police record involving serious crimes. Today every volunteer's past is closely examined. But some secrets are still permitted, and the Legion will protect its soldiers from ex-wives or girlfriends, or even police enquiries, unless the offence is serious. This is why there are many serving Legionnaires who enlisted under a false name and whose true identity remains secret.

The unofficial motto 'The Legion Is Our Country' reflects the fact that all those volunteers who are not French by birth accept that loyalty to the Legion has taken the place of loyalty to their own country. Indeed the Legion also becomes a man's home, at least for the five years of his contract, though the great majority of Legionnaires serve for twenty or more years.

In return, the Legion provides a man with a uniform, a weapon, food and pay. Extraordinarily, it also provides a brothel, and throughout the Legion's long history there have always been more volunteer whores than places for them. Today, if the Legion is sent overseas to fight in a protracted war, then a number of whores volunteer to follow the soldiers. These women are provided with food, shelter and constant medical supervision, but in return they are allowed to service Legionnaires only. A surprising number of them end up marrying Legionnaires.

But it is the spirit instilled by the Legion through training and camaraderie that makes these men become such great soldiers. The officers and NCOs nurture in their recruits a remarkable *esprit de corps,* a conviction that they are special men who would risk their lives for one another and for the Legion. It is true that Legionnaires have fought to the death when a battle is lost and survival impossible, not only for their fellow

Legionnaires but simply for the honour of the Legion. The Legion's motto sums up the character of the unit: 'Honour and Loyalty.'

From the start of their training, would-be Legionnaires accept that the Legion fights and dies but never surrenders. And there is no other known Special Forces unit which can match that extraordinary commitment. Every single Legionnaire has to be very tough, because the physical demands are exceptional. Even today the Legion demands that its volunteers are strong, superbly fit and capable of marching an incredible forty miles a day in full kit over any terrain and in any weather. This march is not a one-off training exercise but a volunteer must repeat it many times a year, every year, once he has been accepted as a fully fledged member of the Legion. There is no Special Forces unit in today's armies that demands such a strenuous ordeal, mainly because other crack forces now use helicopters, trucks and jeeps for speed and mobility.

Today the French Foreign Legion numbers some eight thousand men, though at times in the past it boasted as many as fifty thousand fully fledged Legionnaires. Each year, some six thousand men, mostly ex-soldiers, apply to join the Legion, and less than a thousand are accepted. Most of the recruits are French, but there are also Germans, British, Americans and a number from the former Yugoslavia.

The selection process takes three weeks, which gives the Legion time to carry out physical, psychological and IQ tests as well as to check on the background of the volunteers. If tests show that a recruit is an alcoholic or drug addict he will be dismissed immediately. The language of the Legion is, of course, French and from the start of training the recruit begins to learn the language. He knows he must strive to become fluent, for his life might one day depend on it. After signing on for five years the recruit will hand over all his own clothes, which are taken away and sold. In return he will receive his *paquetage* – tunic, boots, green tie, beret and the famed white cap, or kepi.

A sixteen-week basic training course follows. Today the training is not as brutal and harsh as in former times – recruits are no longer flogged – but still extremely tough. Unable to stomach the training or the discipline, many recruits desert and flee abroad, never to be seen again. Those who stay the course know that the first sixteen weeks is only basic training and that after passing this stage they must continue training to obtain the objective – the highest standard of any fighting unit in the world.

One of the most courageous battles ever fought by the French Foreign Legion in its entire history was at Dien Bien Phu in North Vietnam in the spring of 1954. Vietnam was then part of the French colonial empire in Indo-China, but was coming under increasing attacks from the Vietminh peasant army led by the famous Ho Chi Minh and the brilliant guerrilla leader General Giap. Financed and equipped by the might of Mao Zedong's China, the Vietminh leaders were determined to push the French out of Indo-China for good.

The battle of Dien Bien Phu was the final straw for the French in Indo-China, though they had been fighting sporadic battles against the Vietminh since 1948. Many Legionnaires had earned their spurs in Vietnam confronting this army which relied on machetes, rifles, bayonets and the odd machine gun. By contrast, the French had not only rifles but also machine guns, grenades, artillery, mortar and war planes to keep the enemy at bay. It was an unfair fight despite the fact that General Giap's army was some ten times larger than the French forces and he could always call on tens of thousands of extra troops whenever he wanted.

During the winter of 1953–4 the French generals had devised

a plan to bring about a swift, once-and-for-all victory against the Vietminh. They believed that the only reason they had not wiped out the enemy during the previous six years was the fact that they would never stand and fight. The tactics of the Vietminh were always those of a guerrilla army – sneak attacks against outposts and small units of infantry – followed by a quick retreat into the surrounding jungle. Almost the whole of North Vietnam was a mountainous jungle and therefore no major pitched battles had ever taken place. A few set-piece battles had occurred in which the smaller French forces had devastated vast numbers of the enemy with comparative ease. French attempts at bombing and shelling had proved fruitless because every Vietminh soldier carried, in addition to his rifle, a spade with which to dig deep trenches where he could shelter from bombs and shells.

Nevertheless, the French plan was to lure General Giap and his army into the vast and beautiful jungle valley of Dien Bien Phu, north of Hanoi, close to the Chinese border, and then slaughter them in their thousands by non-stop bombing and shelling. Long enough to land transport planes and wide enough to defend with troops and artillery, the valley rose on either side to high mountains covered with lush green vegetation.

If they could wipe out Giap's huge force at Dien Bien Phu, the French generals reasoned, they would be able to control the so-called Ho Chi Minh trail – the jungle route through which the enemy ferried arms and ammunition to their forces fighting in South Vietnam. In time, they believed, Ho Chi Minh would be forced to the conference table and the whole war in Indo-China would be brought to a respectable conclusion with no further loss of French blood. It seemed a brilliant plan.

In February and March 1954 the French poured some sixteen thousand men, mainly paratroopers and Legionnaires, into Dien Bien Phu. The first task was to flush out and kill Vietminh troops hiding in trenches on the jungle slopes either side of the valley. They then prepared a defensive ring of trenches and fortifications, gun emplacements and strongholds around the lengthened airstrip. Protecting this would be vital to the success of the battle plan, as the French troops, some seven hundred air miles from their bases in the south, would have to be re-supplied by air. There were, in all, seven of these fortified positions, each given a French woman's name.

When all this had been completed the French troops sat around and waited for the Vietminh to attack. Fifty per cent of them were Legionnaires, impatiently waiting for the battle to begin. Some twenty per cent were French paratroopers and another twenty per cent were hardened French infantry. In addition there was support from artillery and engineers.

General Giap obligingly played the role the French had carefully prepared for him. He gathered together an army of fifty thousand hardened Vietminh fighters with which he intended to surround and then wipe out the entire French force at Dien Bien Phu. It was the opportunity for which he too had been waiting and planning.

But this time there was a major difference. During that winter he had persuaded Mao Zedong to provide hundreds of heavy artillery weapons and rapid-firing anti-aircraft guns, as well as trained Chinese soldiers to fire and maintain them. Never before had the Vietminh had such sophisticated weaponry, for they had always relied on guerrilla tactics. Now, for the first time, they would have the weapons to tackle the French.

At dawn on March 13 1954, the French were taken completely by surprise when the Vietminh launched a devastating heavy artillery barrage from the hills surrounding the valley. The shells came whistling down in their hundreds on the French positions and along the entire length of the airstrip. What amazed the French generals was not only that the

Vietminh possessed such devastating, modern artillery but that they had somehow managed to pull these huge artillery pieces up the slopes of the hills in such numbers – a feat they considered virtually impossible, given the mass of vegetation and jungle. And they had done so without the French having any idea what was going on.

The first onslaught was on fort Béatrice to the north of the valley, guarded by one of the toughest Legionnaire brigades. For hours the Vietminh rained down shells on the blighted Legionnaires, who could only take shelter from the bombardment. The French artillery was shelled to pieces and their hand weapons were useless in such circumstances. After five hours of horrendous shelling there was a respite of perhaps a minute and then the Legionnaires heard the wails and screams of thousands of rampaging Vietminh soldiers racing out of the jungle towards their dugouts. The Legionnaires were waiting for them.

The Vietminh, dressed in their pyjama-style fatigues, flung themselves on the barbed wire surrounding the fort. Those who managed to survive the Legion's withering automatic fire hurled grenades into the dugouts before scurrying away to safety, chased by a hail of bullets. Hundreds were killed in that first onslaught, scores left hanging lifeless across the barbed-wire fence.

As soon as the Vietminh had pulled back into the jungle, mortar shells began raining down on the Legionnaires as they went about tending their wounded, re-supplying their ammunition and checking their arms. For the following six hours the mortars and shells kept up a non-stop bombardment of Béatrice, destroying most of the defences around the fort. Shortly after midnight the guns fell silent and immediately the Vietminh infantry swarmed down from the slopes towards Béatrice. Once again the Legionnaires were waiting for them, but there were only about a thousand Legionnaires and infantrymen at Béatrice whereas the invasion force was at least ten times as large. The Vietminh would rush towards the French defences in waves, blowing bugles before hurling grenades into the dugouts and throwing themselves on top of the defending soldiers. With their defences overrun, the order was given for the Legionnaires to pull back to another fort and surrender Béatrice. Twenty minutes after the first attack, all firing stopped. Only about two hundred Legionnaires made it to safety; the rest had been shot, hit by artillery or hacked to death in the trenches.

The following night the Vietminh artillery opened up on the next fort, Gabrielle, and four hours later some six thousand Vietminh swarmed towards the structure on three sides, shouting and screaming and firing their weapons. Fewer than four hundred French troops were defending this fort. There were, however, some small French tanks, and these kept up a steady fire, cutting swathes through the lines of attackers. The first assault was finally repulsed and the small Legion tank force, supported by Legionnaires, counter-attacked, driving away the Vietminh, who turned and fled. But the damage to Gabrielle was irreparable and the decision was taken to abandon the fort and pull back to other positions. In taking Béatrice and Gabrielle, the Vietminh had killed about a thousand Legionnaires, but had lost more than three thousand men.

French bombers flew over the valley intent on blanket-bombing the Vietminh, forcing them to pull away from the airstrip. But, to their surprise, they were met by scores of anti-aircraft guns firing non-stop. The bombers were forced to pull away after taking many hits and the accompanying fighter planes had little success against the enemy guns, as these were well concealed in the jungle. This was the first time in the Indo-China war that the Vietminh had ever used anti-aircraft guns, and it completely altered the French battle plan. At a stroke the French had been forced on to the defensive, for they had neither bombers to take out the enemy hordes nor fighters to destroy the enemy guns, and re-supplying the garrison with troops, ammunition, food and medicines was all but impossible. The Vietminh's use of anti-aircraft guns also meant that sending in paratroopers was now a risky option, both for the paratroopers themselves and the transport planes.

Forty-eight hours after taking Gabrielle, the Vietminh moved on to fort Anne-Marie, but this time they changed tactics. They didn't send in any ground troops after hours of bombing, but for three days and three nights kept up almost non-stop shelling of the fort, making normal life impossible for the Legionnaires holed up inside. The decision was taken to vacate yet another fort and pull back to reinforce the airstrip defences.

But the airstrip had taken a tremendous pounding during that first week of the battle and was declared unusable. All supplies would now have to be dropped by parachute or helicopter. But, given the new-found firepower of the Vietminh, re-supplying by chopper was a very dangerous manoeuvre. So accurate had the Vietminh gunners become, putting many transport planes out of action, that parachute drops during daylight were abandoned. But dropping supplies in darkness was always a risky option. The French battle plan at Dien Bien Phu was rapidly becoming a nightmare and the Vietminh were creeping ever closer.

By day the Vietminh continued their barrage of the main French defences surrounding the airstrip. At night they adopted a new tactic, crawling as close as possible to the perimeter wire around the airstrip and then digging like crazy with their spades. The French replied with mortars, killing many of the thousands of Vietminh peasants digging their trenches around the airstrip. By day the French filled in the trenches, but the Vietminh were crawling ever closer.

For two weeks this stalemate continued, but fewer parachute drops were hitting the target as the Vietminh encircled the French troops and supplies of ammunition, food and medicine were getting lower. It was impossible to take out the wounded, and doctors and surgeons brought in by helicopter were having great difficulty keeping the seriously injured alive.

The French decided to reinforce their men at Dien Bien Phu and two parachute battalions, numbering some fifteen hundred men, were successfully dropped one night as Legion troops took up positions around the perimeter of the airstrip in case the Vietminh tried to take out the paratroopers.

This influx of new troops restored morale, but two weeks later talk began of beating a defensive retreat out of the valley, though, in reality, everyone at Dien Bien Phu knew that they would be slaughtered on the way out. And no one was prepared to leave behind the wounded and the dying to be massacred by the Vietminh.

When pulling back was mentioned, however, the Legionnaires refused to even discuss the matter. They were adamant. They replied, 'We are not leaving a single Legionnaire here to be massacred by the enemy... We fight to the death if necessary... we prefer to die rather than retreat... there is no pulling back... there is no surrender.'

Each day the French were sending in their bombers and fighter planes but to little or no effect despite the fact that large chunks of the surrounding jungle were being set alight and decimated by the napalm bombs. This reliance on bombing was meant to be the core of the French offensive. It had failed completely. The Vietminh continued to press forward at night and to shell the French compound throughout the day.

As word spread of the impending disaster at Dien Bien Phu those Legionnaires in other parts of Indo-China reacted by insisting on going to join their comrades trapped in the valley.

There was now no possibility of these Legionnaires parachuting into the compound and so they made their way on foot through the jungle. When they reached the perimeter, which was guarded by the surrounding Vietminh forces, they would surprise them, shoot their way through and join their comrades. It was foolhardy and reckless, but there was no way the French commanders could stop the Legionnaires.

By mid-April the situation had become critical. For one month the defenders had been encircled, somehow withstanding the shells and the mortar bombs which peppered the defences every hour of daylight. At night they could now hear the Vietminh attackers tunnelling away with their spades, getting ever closer.

Finally, on May 1, General Giap decided the time had come for the final push to overrun the French positions and take all the forts and the headquarters. In the early hours of the morning, in driving rain, the Vietminh bugles were heard once more and the defenders prepared for yet another onslaught. But this time it would be different.

Coming at them out of the darkness, the French saw thousands of enemy soldiers yelling, firing their rifles and running flat out towards their positions. Never before had they experienced such an infantry attack – some seven thousand troops were attacking in wave after wave, pressing towards the remaining forts around the airstrip. But, as before, the French were waiting for them, their machine guns rattling away non-stop, carving huge swathes through the Vietminh lines. Even so, as the Legionnaires felled one wave of attackers, another wave would take their place, running in their bare feet through the torrential rain towards the French lines, screaming above the din of battle.

Repeatedly, lines of French defenders were overrun by sheer numbers and forced to pull back. Legionnaires held in reserve were sent in to counter-attack and every time their onslaughts drove the Vietminh back. At various points in the defence this happened three and four times during the five-hour battle as the horrendous pressure was kept up by General Giap's forces. It was painfully obvious to the French officers that defeat was at hand, as Giap clearly cared nothing for his troops, sending them on such certain suicide attacks in which hundreds of them were cut down as they ran straight towards the enemy machine guns.

As dawn broke the Vietminh beat a retreat to safety of the jungle, leaving some fifteen hundred dead comrades in or around the French positions. In places their bodies were heaped one on top of the other and had been used by the French as defensive cover. In contrast, the French had lost about two hundred men.

Yet there was no respite for the defenders. No sooner had the Vietminh retreated than their artillery started up again, pounding the French defences. Two nights later General Giap sent in fresh troops in a bid to finish the battle. The tactics were the same – wave after wave of attackers flooded through the shattered perimeter wire in the dark hours of the morning intent on overrunning the French defences with weight of numbers.

Against all the odds, the French somehow held out, with the Legionnaires relying on fearsome counter-attacks to dislodge the enemy. But ammunition was now running dangerously low, the wounded were lying unattended in the dugouts and the French were taking a mighty beating.

But the Vietminh were suffering too. Thousands were lying dead on the airstrip, thousands more were wounded, and the peasant army seemed almost on the brink of cracking under the skill and courage of the Legionnaires. For four days there were no ground attacks, but there was still the ever-present shelling, which made life hell for those desperately trying to tend the wounded, repair the damage, eat a meal and take some rest. The defenders were physically shattered and close to exhaustion.

Colonel de Castries, in command of Dien Bien Phu, knew his men could take no more and prepared to surrender the garrison. He called French headquarters to report the desperate plight of the defenders and the necessity to surrender to save the lives of some of the men under his command. He was given permission to do so. But that was not the end of the battle. The Legion's paratroop commanders heard that Colonel Castries was about to order the white flag of surrender to be hoisted and sent a runner to him with an extraordinary message:

'The Legion never surrenders... We're going to attack.'

Two hundred Legionnaires fixed bayonets, charged the magazines of their light machine guns and sub-machine guns and clambered out of their defensive positions. They began walking towards the enemy beyond the perimeter, holding their guns at their hips, fingers on the trigger. They then began to trot line abreast and as the Vietminh began firing at the human line moving towards them the Legionnaires charged, yelling, '*Faire Camerone*', or 'Let's do a Camerone.' This battle cry harks back to the Legion's most famous stand, when just fifty Legionnaires, holed up in a farmyard, held at bay an entire Mexican army of some four thousand soldiers, including cavalry. When only five Legionnaires were still standing and capable of running, they fixed bayonets, raced out of the farmyard and towards the astonished Mexicans, yelling obscenities. It was only when the Legionnaires were some twenty yards from the enemy lines that a whole company of Mexicans opened fire, killing all five of them. It was the most glorious battle in the French Foreign Legion's long and distinguished history.

The French Foreign Legion was formed in 1831 for the sole purpose of getting rid of thousands of soldiers from across Europe who had moved to France after the collapse of Napoleon's empire. One thousand soldiers, ranging from teenagers to sixty-year-old veterans who had served in a number of European armies and wanted to return to army life, were hurriedly mobilised and the ragtag Legion was born. However, all the officers recruited were French. The Legion's first campaign, soon after its formation, was in Algeria, which would become its natural home, away from the politics of France.

Throughout the following decades recruits to the Legion came from all parts of Europe and, according to legend, many joined to escape justice, wives, girlfriends and money lenders. In fact many were simply refugees or former soldiers who had found life too tough outside the army. After every major war in Europe during the past one hundred and fifty years the Legion has been a haven for veterans from many countries wishing to continue army life. During those years many recruits have been French, German, Spanish and Swiss, with some ten per cent British.

The Legionnaires found life very tough in Algeria, where they were engaged in a war of colonisation against tough tribesmen who fought guerrilla-style battles against the ill-prepared and poorly armed Legion.

The Algerian Arabs relied on spear-wielding cavalry and the Legion's only defence was to form into squares and fire their muskets at the horsemen as they galloped around these formations. Hand-to-hand fighting usually ensued in which the experienced Legionnaires dominated, driving away the Arabs who weren't killed on the field of battle.

Any captured Legionnaires were handed over to the Algerian women, who would take a delight in stripping their captives naked and indulging their whims with exquisite methods of torture which always ended in terrifying pain before the Legionnaire's eventual death. The women would then hurl their captives' testicles and severed heads at Legion troops passing through their villages.

The conquest of Algeria took four years, in which time the Legion became a tough fighting force. It met most of its own needs, relying on France only for weapons, ammunition and the paltry pay. Those years produced a tightly knit band of disciplined brothers who would die for one another rather than surrender to an enemy. The Legionnaires also learnt that there was no question of surrender, because death would follow automatically and sometimes in the most gruesome fashion. And slowly, through the nineteenth and twentieth centuries, the Legion's reputation grew, and those who served in it were treated with respect and often a little fear.

Today the French Foreign Legion consists of some eight thousand soldiers, mainly infantry and paratroopers, but it also includes a mountain division, artillery and engineers. A rapid-reaction force of some fifteen hundred men, all with Special Forces skills, is based in France, and there are also Legionnaires based in Chad, Surinam, Djibouti, French Guinea and the Comoros Islands, near Madagascar. Some six thousand men apply to join the Legion each year, but only about a thousand are accepted.

One of the Legion's more recent and most dramatic operations was the rescue, in 1978, of some two and a half thousand European men, women and children in the copper-mining town of Kolwezi, in Zaire, after two thousand heavily armed rebels of the Congolese National Liberation Front stormed the town in a bid to steal arms and money.

These wayward rebels arrived in armoured cars in Kolwezi, and, with much firing of arms, took control of this town of some twelve thousand people. The local Zairean militia of two hundred armed soldiers offered no resistance and quickly fled, leaving the rebels in complete control. Their first target was the bars and hotels, which they shot up before drinking the town dry and then embarking on an orgy of drunkenness and violence, raping the women and shooting any man who dared to get in their way.

The rebels set up courts in the hotels and hauled local dignitaries before them, accusing them of aiding President Mobutu of Zaire, a man they described as a traitor. Every person brought before the court was found guilty and all were immediately shot. The orgy of violence and drunkenness continued. Fortunately, however, a radio operator at one of the mines managed to get a message out telling the civilised world what was going on and pleading for assistance to stop the murder and violence.

The Zairean government was not capable of putting together a military force strong enough to take on the rebels, and Belgium, the former colonial power, refused to intervene. The French came to the rescue by sending in Legionnaires from the 2nd Parachute Regiment from their base in Corsica. Within ten hours the soldiers were flying to Kinshasa, but the following day, when they were over Zaire and ready to parachute into Kolwezi, they faced major problems. They had no maps of the town, no information about the size or capability of the rebel forces and no idea where the town's two thousand three hundred white men, women and children might be holed up. Nor did they have artillery or mortars – only grenades, rifles, sub-machine guns and light machine guns. The United States sent five Hercules C-130 transport planes to the Zairean capital, Kinshasa, to pick up the Legionnaires as they parachuted down there instead, and within hours the first batch of four hundred Legionnaires were on their way to Kolwezi. They had no idea they would be facing a rebel army of some two thousand men with mortars, medium machine guns and armoured cars.

Their orders were to make straight for the town centre because it was believed that the rebels would have herded the town's whole population into one place. They had been warned that they would have to move at great speed because this group of Congolese rebels was renowned for killing hostages as soon as trouble arose.

The scene that greeted the Legionnaires as they entered the outskirts of Kolwezi was horrendous. Scores of swollen, decomposed bodies, both black and white, were lying around the streets, in alleys and doorways. Even young children had been massacred, their frail, thin bodies left to rot in the scorching sun, and the stench of rotting corpses filled the air.

As the Legionnaires pushed on towards the town centre the resistance became tougher, because the rebels opened up with mortars and machine guns. But the Legionnaires were taking no prisoners. They had seen that the rebels had shown their victims no mercy and they were determined to find and kill every rebel they came across. As the townspeople heard the sound of gunfire they threw caution to the wind, grabbed their children and ran towards the sound of the shooting in the hope of finding safety. This caused problems for the Legionnaires as they took aim at the rebel troops. And the fleeing women and children also provided the rebels with easy targets for their machine-gunners, who mowed down twenty or more before they reached the safety of the Legionnaires.

The townspeople were able to give valuable information to the Legion, telling them that the rebel command had taken up positions in the principal hotel and the main police station and that marauding gangs of rebels were roaming the streets, mostly drunk, and killing blacks and whites at random. They also reported that some of the rebels seemed to take great delight in torturing and killing the whites, making them dance in the street while they fired shots at their feet. When, through exhaustion, the victims could dance no longer, the rebels shot them in the head, while roaring with laughter.

The Legion knew they had to take command of the town that day and somehow hang on against overwhelming odds because the next batch of four hundred Legion paratroopers would not arrive until the next morning. The decision was taken to secure both the hotel and the police station, which could then be defended through the night. During the afternoon sections of paratroopers went into the Old Town to tackle the rebel gunmen, who could be heard firing intermittently. Firefights took place, but the rebels were no match for the accurate, disciplined Legionnaires, who had no compunction in shooting to kill every armed rebel they came across that day.

There was only one main attack which seriously tested the Legionnaires metal, and that was when the rebels took their three armoured cars into the Old Town, firing their powerful machine guns from the turrets. Behind the armoured cars came rebels who also kept up a steady stream of fire, pinning down the Legionnaires in the alleys and shanty buildings. The Legionnaires replied with grenades, which they hurled into the melee of rebel troops sheltering behind the armoured cars. It worked, and after a few minutes the rebels turned and scattered while the Legionnaires opened fire, killing some eighty of them. Fearing they might be cut off, the armoured cars withdrew and the Legionnaires went about their task of searching out armed rebels and killing them on the spot.

But the main thrust of the Legion's first day in Kolwezi was to take the two main objectives – the hotel and the police station. First, they targeted the hotel. One platoon of Legionnaires was sent to the right flank, and when in position they opened fire on the rebels defending the building. As soon as the rebels rushed to fire at the flanking soldiers, two platoons of Legionnaires made courageous frontal assaults on the hotel, racing in firing sub-machine guns from the hip and causing so much alarm and mayhem in the hotel that the defenders simply fled rather than take on the charging paratroopers.

The police station proved more problematic because it was easier to defend. Here again the Legion officers decided on a frontal assault, but only after a flanking platoon had the defenders holed up with constant sniper fire. As dusk fell the Legionnaires once again raced towards the front of the police station, firing as they went, while those on the flank opened up with machine guns, forcing the rebels inside to lie low. Within sixty seconds the Legionnaires had reached the walls of the police station and they lobbed grenades through every window. In a bid to save themselves, a dozen rebels sprinted out of the building, but they barely made three or four yards before the Legionnaires gunned them down.

Now the Legion had two positions which could, with luck, be defended until relief arrived the next morning. But it would prove a long, hot night, for the rebel commanders repeatedly tried to recapture both buildings. At one point both the hotel and police station were surrounded by a few hundred rebels who fired non-stop for thirty minutes. But the Legionnaires handled the attacks with confidence, picking off a number of rebels, which seemed to sap the will of the others to take further action.

As the arrival of the other parachute battalion drew near, the Legionnaires turned up the heat, taking the fight to the weary and hung-over rebels. They wanted to make sure that their comrades dropping out of the skies would not receive a hostile reception. They also gambled on the fact that the Congolese rebels would not relish tackling two battalions of Legionnaires, one at their front, the other at their back. And they were right. One minute the rebels could be seen moving out of town towards the airport, and the next they seemed to be dispersing in every direction.

Legion jeeps and trucks also arrived that day, but as the

second flight of four hundred paratroopers moved towards Kolwezi to link up with their comrades, the Congolese rebels decided the time had come to quit the town and escape back to the safety of Angola, from where they had come just a few days before. On their arrival the battalion had no intention of letting such butchers off the hook so lightly and pursued them relentlessly right to the Angolan border. They were shown no mercy. Some hundred or so rebels were slaughtered during their retreat.

Undeniably, the rescue of more than two thousand white men, women and children in Kolwezi was a brilliant and courageous Legion success, though they were unhappy that some two hundred and fifty whites and some two hundred blacks had been slaughtered in the most appalling circumstances by the rampaging rebels. In the entire operation the Legion suffered five dead and twenty-five wounded.

This author was with the French Foreign Legion in Lebanon in the summer of 1982, when the United Nations called on the Legion to escort the Palestinian leader, Yasser Arafat, and his guerrilla fighters to safety. Arafat and his thousand-strong force were retreating in some disarray before the Syrian army, which was hell-bent on destroying his Palestinian fighters.

I was with Arafat's headquarters unit, made up of six of his lieutenants and advisers and some twelve heavily armed bodyguards, retreating towards the safety of Tripoli, when the pursuing Syrian forces caught up with the rearguard. We were holed up in a cave outside Tripoli, with the Syrians firing machine guns at our position, when a platoon of some thirty Legionnaires arrived on the scene. They took control of the immediate area, fanning out in a show of strength and warning the Syrians to stop firing and back away, otherwise they would launch an attack to remove them. While the Legionnaires held the position facing the Syrians, Arafat and his men slipped away and reached the safety of a house in Tripoli which had been secured by the Legion. Without a further shot being fired, the Syrians faded away. Some ten hours later I watched Arafat and his men set sail for the safety of Libya as the Legion held the port.

CHAPTER 4
COUNTERING TERROR

THE LATE 1960S AND 1970S saw a dramatic upsurge in terrorist activity in various parts of the world, particularly Europe and the Middle East, and during this period the man who became known as 'Carlos the Jackal' was the epitome of the international terrorist. Mysterious, cunning, courageous, ruthless and intelligent, he was one of the most effective and sure-footed terrorists ever known. Carlos was born Ilich Ramirez Sanchez in 1949, the son of a Venezuelan millionaire lawyer who openly trumpeted Marxism and inspired his son to follow his fervent commitment to the communist cause. As a teenager, Carlos joined various Venezuelan terror groups opposed to the nation's dictator, President Raul Leoni, and took part in a number of low-key urban guerrilla operations.

After being educated in London and Paris, Carlos spent a year at the Patrice Lumumba University in Moscow, where he was indoctrinated still further in the ideology of communism. He went on to join the Popular Front for the Liberation of Palestine (PFLP) in Beirut, which was where he was given the nickname 'Carlos'. Under the tuition of the Palestine terror groups he trained to become a first-rate terrorist, expert in weapons and explosives and guerrilla tactics. The PFLP leadership, however, quickly came to realise that Carlos was no ordinary foot soldier but a highly intelligent young man who could be trained to become an outstanding strategist. From Beirut, he was sent to London, where he was tasked with selecting potential targets for assassination or kidnapping. He drew up a list of some five hundred people who then dominated the worlds of politics, business and even the arts.

In October 1973 a brief war erupted in the Middle East, but before Israel could press home a retaliatory victory against Egypt and Syria, the Soviet Union demanded that the United States bring the conflict to an end, otherwise it might have to intervene to prevent the aggressors, who were two of its Middle East client states, being crushed. A few months earlier, following the death of one of the PFLP's senior planners, Carlos had been activated as its chief assassin, and now, in the wake of the October war, he carried out the first of his many terror attacks. He arrived at the London home of Edward Sieff, the chairman of Marks & Spencer, and when the butler opened the door he held a revolver to his chest, ordering him to take him to his employer. They walked into the bathroom, where Carlos shot Sieff in the face, turned on his heel and ran. But somehow Sieff survived the point-blank attack. The following day the PFLP claimed responsibility for the shooting.

Within weeks Carlos went into action again, throwing a bomb into the City of London branch of an Israeli bank. The bomb failed to go off but a witness to the attack gave a description of a suspect that matched Carlos in every detail. By now London had become too hot for the terrorist, so he moved to Paris and carried out bomb attacks on three pro-Israeli newspapers. He also bombed a crowded store in the city, killing two people and wounding more than thirty. Days later he shot dead the military attaché of the Uruguay Embassy in Paris.

Rapidly, Carlos was earning himself a formidable reputation, particularly among the security services of Israel, the United States and all western Europe's democratic nations. His one-man terror campaign was also proving highly embarrassing to governments whose responsibility it was to track down and arrest or kill him. But a stop had to be put to his daring, pitiless violence.

Aided by a single accomplice, Carlos opened fire with a Soviet anti-tank rocket-launcher on an Israeli El Al Boeing 707 carrying one hundred and thirty-six passengers and crew as it taxied slowly along the runway at Orly airport in Paris in January 1975. The terrorists' first rocket missed the target but hit a Yugoslav Airlines DC-9; their second rocket also missed. In panic the two men fled the scene and the launcher was later found on the back seat of a car stolen earlier that day. A week later Carlos returned to Orly accompanied by three Palestinians, but they were spotted and fled. He melted away in the crowd while the three Palestinians grabbed some hostages and dragged them into a lavatory. Eventually the hostages were released unharmed and in return the three gunmen were given safe passage to Baghdad.

Then, in December of that year, Carlos led a group of six terrorists, including one woman, into the headquarters in Vienna of the Organisation of Petroleum Exporting Countries, where the OPEC oil ministers were meeting. This well-planned attack would make Carlos the world's number-one terrorist as television pictures of the incident were flashed around the world.

The group walked into the first-floor offices firing automatic machine pistols, shot dead three people, including one police officer, and stormed into the room where the oil ministers were meeting. Shouting and yelling at everyone to stay still and not move, they rounded up all eleven ministers and fifty members of staff.

As armed police rushed towards the building the terrorists took up positions at the windows, firing at the officers below. One terrorist and a police officer were wounded. The group freed a woman hostage and, along with her and the injured terrorist, they sent a ransom note. In this they demanded that a political communiqué written by Carlos should be broadcast on Austrian state radio and that an aircraft should be made available to fly the terrorists and the OPEC ministers and the other hostages to the Middle East.

The Austrian authorities had little option but to agree to the demands and the terrorists and all their hostages were driven to Vienna airport in a bus with blacked-out windows. Television pictures showed the gun-toting terrorists on the tarmac ushering their hostages on to the Austrian Airlines DC-9. Only when airborne did Carlos tell the pilot that he must fly to Algiers. Shortly after the plane landed, all but fifteen hostages were released and the terrorists took another aircraft to neighbouring Libya.

In Libya, the terrorists made further demands. They wanted a huge ransom from Saudi Arabia and Iran in return for the lives of their respective oil ministers. After phone calls between King Khaled of Saudi Arabia and the Shah of Iran a deal was agreed and a ransom of some forty million dollars was transferred from Switzerland into the account, in a bank in Aden, of the PFLP's Special Operations Group. The following day the aircraft, with some of the hostages still on board, returned to Algiers. After confirmation that the ransom had been deposited, all the hostages were released and Carlos and his accomplices were driven away, bringing to an end one of the most public, audacious and successful terrorist actions of modern times. Carlos was now the world's most wanted terrorist.

In June 1976 Carlos, working with Wadi Haddad, the PFLP's head of terrorism, masterminded the hijacking of an Air France Airbus from Tel Aviv to Paris via Athens. This hijacking would result in one of the most dramatic and courageous rescue missions in the history of modern terrorism, an operation which drew praise from Special Forces soldiers throughout the world. The story of the raid at Entebbe is told in the following chapter.

Carlos did not take an active part in the Air France hijack, nor the action at Entebbe. But, funded by millions of dollars from the OPEC ransom, he went on to set up his own terrorist action group, recruit his own elite force of hijackers and gunmen from around the world and, as a safeguard, work in co-operation with the Mukhabarat, the highly secretive Iraqi intelligence agency. Between 1979 and 1981 he and his terrorist group were recruited by Colonel Gadaffi of Libya to track down his opponents who had fled in fear from Libya, and assassinate them. In return Gadaffi would pay handsomely for the various operations and supply Carlos with all the arms, ammunition and terrorist supplies he required. When necessary, aircraft, ships and boats were provided, as well as access to safe houses in a number of European countries. Western intelligence agencies confirm that during these two years a number of political opponents of Gadaffi who had fled Libya were assassinated in mysterious circumstances. However, Carlos came to consider the Libyans unprofessional and inefficient, and decided to offer his services to other terrorist states.

In the early years of terrorist activity by Carlos, the PFLP and Abu Nidal in the Middle East and Europe, the IRA in Northern Ireland, the Baader-Meinhof Gang in Germany, Direct Action in France, the Red Brigades in Italy and ETA in Spain, most countries relied on police forces and one or two Special Forces to track down and take out those who were prepared to hijack and demand ransoms. But that attitude changed dramatically as terrorist groups recruited more violent volunteers to their cause. Not only were the numbers of Special Forces members greatly increased but, in some cases, regular soldiers also had to be deployed in a bid to either bring terrorists to justice or kill them.

Back in the late 1960s the Secret Intelligence Services of most western European nations and the United States had been warning their governments of the increase in covert terrorist cells seeking to subvert or even replace democratically elected governments by the use of violence, murder, kidnapping and bombing. Many of this new breed of terrorists were bright, intelligent, middle-class and well-educated young people who felt they had a mission in life. Some believed that they were correcting legitimate wrongs, and demanded that minorities be given the same privileges as the majority. Others, with some justification, were supporting civil rights in various European countries, inspired by the success of such campaigns by blacks and Hispanics in the United States.

What disturbed western European intelligence services was that some of the people they were keeping under surveillance were prepared to go to any lengths, including the slaughter of innocent civilians, to push forward their demands, which were usually hard-left or communist-inspired. But because intelligence staff were rarely able to infiltrate these secret terrorist cells, they could only react to terrorism after the event; they were powerless to prevent atrocities taking place. Neither the standing armies of Europe nor the civilian police were trained or equipped to carry out successful anti-terrorist activities, which meant that governments needed to recruit, train and deploy squads of tough, ruthless paramilitary soldiers in a bid to halt the terrorists' war on democratic governments.

Governments became convinced that the only way to combat terrorism was to meet force with force. They also agreed to co-operate much more closely on sharing intelligence about terrorist groups despite the fact that many of these organisations were concerned with their own national issues rather than the broader aspect of undermining European democracy as a whole. That would come later, when terrorist groups began swapping intelligence and tactics in an effort to defeat the forces of law and order across national borders.

In establishing or revamping their Special Forces, most European countries took as their template the British SAS, which had already shown its worth in counter-terrorism operations in various parts of the world outside Europe since World War Two. It was fortunate that they did so, for terrorism was set to create serious problems for many European democracies during the coming decades. Indeed the continuing difficulties facing governments and their Special Forces were underlined by the events of September 11 2001 and the subsequent global war on terrorism urged on its allies by the United States.

As early as 1972, West Germany set up the Grenzschutzgruppe-9 (Border Guard Group 9), known as GSG-9, under the direction of Colonel Ulrich Wegener, a former FBI-trained police officer. Wegener was a highly motivated, disciplined leader who recognised that the volunteers for this new unit had to be remarkable young men. Sixty per cent of volunteers were turned down and those who passed the tough examinations underwent six months of basic training and three months of advanced work.

Wegener laid down that the basic skills must include unarmed combat, marksmanship, martial arts and, somewhat surprisingly, the study of German law. The advanced stage included training for assaults on buildings, aircraft and trains, close-quarter combat and high-speed driving, the last practised in the dead of night on Germany's autobahns at speeds in excess of one hundred and twenty miles per hour.

So keen was Wegener to learn Special Forces skills that he was in constant touch with the British SAS and Israel's Mossad, and was contacted as soon as the Air France jet involved in the

Entebbe raid was hijacked and diverted to Uganda. He flew immediately to Israel and then to Entebbe with the one hundred men of Israel's 35th Parachute Brigade to see for himself how the Israelis planned and carried out the release of the hostages.

One year after recruitment of the initial two hundred men, GSG-9 was operational and working out of St Augustin, near Bonn. There was a headquarters unit and three combat teams, each of thirty men and divided into six five-man sections. GSG-9/1 was a pure combat assault team; GSG-9/2 became a one hundred-man maritime section protecting German offshore installations; and GSG-9/3 became an airborne and paratroop unit. GSG-9 also boasted its own unique intelligence, communications, logistical and engineering sections, as well as its own helicopter unit. The West German government spared no expense in building sophisticated training complexes and aircraft interiors, and supplying speedboats, high-performance cars and helicopters and the very latest in weapons technology and electronic surveillance.

It was not long before the Baader-Meinhof Gang, named after Andreas Baader, the son of a celebrated historian, and Ulrike Meinhof, the daughter of a museum director, began serious terrorist operations. The philosophy behind the group was to create as much violence as possible in the hope of provoking the authorities to overreact and be seen as oppressors. They claimed they existed to represent the oppressed peoples of Europe, to right wrongs, to obtain justice and to establish a fair and just society for all. Their stance gained them support from politicians, the media and academia.

After robbing a bank together, Baader and Meinhof fled to Syria, where they underwent terrorist training. On their return to West Germany they began a reign of terror, robbing a number of banks at gunpoint and stealing some US$500,000, reportedly to finance their acts of terror.

Initially using the name the Red Army Faction and suggesting that their movement had worldwide support, they began their operations in May 1972. Their first attack was the bombing of a US Army base in Frankfurt which killed one officer and wounded thirteen others. Two weeks later they bombed the US Army's headquarters at Heidelberg, killing three and wounding eight. Other bomb attacks took place in quick succession in Augsburg, Munich and Hamburg, as a result of which the death toll reached double figures, with more than sixty injured.

But then the West German police had a stroke of luck. A member of the public tipped them off that a small garage was being used as a bomb factory. Baader and others were arrested, and then Meinhof two weeks later. But the capture of the ringleaders only resulted in more terrorists taking up the cause, and by 1975 it was estimated that there were sixty core members and some two thousand sympathisers. And now the group's tactics changed. Instead of bombing and shooting, the new members turned to kidnapping prominent people and holding them to ransom. On some occasions they demanded the release of their jailed colleagues; on other occasions they demanded money. But the newly formed GSG-9 was never far behind them, ready to be called in whenever Special Forces were needed.

One of the most dramatic and sensational operations carried out by Special Forces in the war against international terrorism involved an GSG-9 assault team and two SAS advisers and followed the hijacking of a Lufthansa Boeing 737 flying from Palma, Majorca, to Frankfurt on October 15 1977.

The four hijackers – two men and two young women – were all members of the PFLP. As the PFLP's terror chief, Wadi Haddad was working closely with George Habash, the group's political figurehead. Together these two men had been organising terrorist activities throughout Europe and the Middle East for ten years, and their ruthless activities were well known to the world's counter-terrorism organisations.

The two male hijackers, the leader Zohair Yousif Akache and Wabil Harb, nonchalantly made their way to the flight deck after the plane had been airborne for thirty minutes. They pushed open the door, drew handguns and appeared to go berserk, screaming at the top of their voices. They yanked the co-pilot, Jürgen Vietor, from his seat and dragged him out of the cockpit. The two young female terrorists, Suhaileh Sayeh and Hind Alameh, also both armed with handguns, pushed the three female cabin crew to the rear of the economy section, where Vietor was made to join them. The first-class passengers were likewise forced into the back of the economy section. As the pilot, Captain Jürgen Schumann, sat at the controls, Harb stood over him with his pistol at his head to prevent him passing a message to any control tower that might have been listening.

Akache stormed up and down the centre aisle, ferociously waving his gun in the air, threatening passengers and hurling abuse at everyone. Some passengers screamed hysterically, others cowered in silence. Then Akache yelled for silence and walked up and down the plane, repeating loudly, 'I am Captain Martyr Mahmoud, and anyone disobeying my orders will be shot.'

All male passengers were ordered from their seats and searched for weapons. The two women then went through all the hand baggage, piece by piece, throwing all the contents on seats in the first-class area. Akache terrified the passengers, particularly the women and children, as he would suddenly scream at someone, hit others with hard blows to the head and face for no apparent reason and berate others, a look of hatred and anger on his face. Some passengers thought he was mentally unbalanced as his outbursts became more hysterical and he repeatedly lost his temper over trivial matters. They feared for their lives.

Four hours later the plane landed at Rome's Leonardo da Vinci airport, where, minutes later, it was surrounded by Italian troops and armoured vehicles. Speaking from the co-pilot's seat, Akache broadcast his demands, which included release the of eleven members of the Red Army Faction from jail in Germany and of two Palestinian terrorists from Turkish jails, and a ransom of US$15 million. Chillingly, he concluded, 'All demands must be met by 8 am tomorrow [Sunday, October 16] or the plane with everyone on it will be blown up.'

The West German Interior Minister urged the Italians to shoot out the aircraft's tyres to prevent it taking off, but they refused, fearful that if they did so the hijackers would blow up the aircraft. Instead the Italian government granted the hijackers' wishes, refuelled the plane and let it fly away. Three hours later, after another horrendous trip for the passengers, the Boeing landed at Larnaca, in Cyprus, and after refuelling took off once more.

Akache demanded Schumann fly first to Beirut, but the Lebanese blocked the runway. Syria, Iraq and Kuwait also refused to let them land. Now the fuel was running low and Schumann told the control tower at Bahrain airport that if they were refused permission to land, the plane would run out of fuel. Still permission was not granted, but Akache ordered Schumann to land there anyway. After touching down, Schumann checked the fuel – three minutes' worth was left in the tanks.

In Germany, GSG-9 had been contacted by the German government and Colonel Ulrich Wegener was ordered to prepare a team to stand by at Cologne airport, ready to fly out to rescue the hostages when their final destination became known. He called the SAS in Britain to request advice and assistance, and within hours the second in command of 22 SAS Regiment, Major Alastair Morrison, along with Staff Sergeant Barry Davies, flew from RAF Brize Norton to Bonn, taking with them the then revolutionary stun grenades nicknamed 'flashbangs', which GSG-9 officers had never before seen. The dazzling burst of light produced by these devices caused a deafening explosion which disoriented and incapacitated people without physically harming them. The GSG-9 officers also took with them an aluminium suitcase containing the full ransom sum in cash.

The GSG-9 team and the two SAS men remained at Bonn for some time as the hijackers flew around the eastern Mediterranean and various Arab countries, becoming increasingly desperate as country after country refused the hijackers permission to land. As it seemed that the hijackers wanted to stay somewhere in that region, it was decided that GSG-9 and the SAS men should fly to Dubai, a good jumping-off point.

Meanwhile conditions on board the hijacked aircraft were becoming horrendous, as the hostages had been forbidden to use the toilets. When the power supply failed and the air conditioning no longer worked, the temperature inside the plane climbed to over one hundred degrees centigrade. Most hostages stripped down to their underwear. Occasionally the hijackers permitted food and water to be brought to the plane, and once permission was given for the toilet tanks to be replaced with clean, empty ones.

Throughout the ordeal Akache underwent dramatic mood swings, sometimes chatting calmly to people and at other times screaming hysterically at passengers and crew, threatening to kill them. At one point he lost control of himself completely. After the hijackers had been given permission to land at Aden, Captain Schumann was allowed to walk to the control tower to plead for fuel from the Yemeni authorities so that he could take off again. When the pilot returned and climbed aboard, Akache went berserk, accusing him of betrayal and waving his handgun at the cool-headed German. Then the terrorist pointed the gun at Schumann's head and shot him in the face at point-blank range.

Afraid that the Yemenis might now decide to storm the plane, Akache ordered the co-pilot, Vietor, to take off immediately and, when they were airborne, told him to fly to Mogadishu, the capital of Somalia, in East Africa. The hostages had now been aboard the hijacked plane for four long days, and many were very agitated and frightened. After Schumann's appalling murder, a number of the hostages were convinced that they would die in one way or another, and some said their last prayers.

Aware of Schumann's death, the West German government feared that the desperate hijackers might blow up the aircraft with all the hostages on board, and gave orders for the rescue attempt to begin. Another Lufthansa plane, with the Special Forces on board, followed the hijacked Boeing 737 to Mogadishu and managed to land without the terrorists realising that another aircraft had arrived. After talks between the West German Chancellor, Helmut Schmidt, and President Siad Barre of Somalia, permission was given for the rescue to proceed.

The assault plan was practised once more on the aircraft that had brought the rescue team. Everything was planned to the last detail. As darkness fell that night, the GSG-9 snipers took up their positions around the hijacked aircraft, but out of sight of those on board.

In the control tower skilled negotiators began a dialogue with Akache, telling him that the West German government agreed to all his demands on condition that all the hostages and crew were released unharmed. Akache was also warned that if anyone on board was harmed, the deal was off. The negotiators told him that the $15 million ransom would be handed over, and that both the Red Army Faction and PFLP members would be released.

Akache, however, countered by refusing to release the hostages until the terrorists had been freed and the ransom money had arrived in Mogadishu. The authorities agreed, telling him that the money had already arrived but that the terrorists, already on their way from jails to the respective airports, would not arrive in Mogadishu until the following morning. This news, given to them by Akache, brought cheers from the hostages.

At midnight the rescue plan began. First, a recce team, with the aid of an image intensifier, crept towards the plane and checked the whereabouts of Akache. They discovered he was chatting with Wabil Harb in the cockpit. As soon as they reported this information the assault team swung into action, moving to a prearranged assembly point behind the plane where no one on board could see them. In single file they moved stealthily towards the rear of the plane and placed a ladder against each wing and another at the rear door.

While this was going on the negotiators informed Akache that the aircraft with the freed terrorists on board had just left Cairo after refuelling and that he would be able to speak to them when it was within range of Mogadishu airport.

At the final briefing the GSG-9 assault team were told, 'Remember, shoot to kill. These hijackers have already shot the pilot dead. We must not risk the lives of any of the hostages. Good luck.'

The assault began at 2 am. Somali troops at the end of the runway lit a huge fire some three hundred yards in front of the plane. This drew the attention of both Akache and Harb. As the flames leapt ever higher, Major Morrison and Staff Sergeant Barry Davies simultaneously hurled their stun grenades at the aircraft. These exploded with tremendous noise and blinding flashes over the wings and cockpit, and immediately the GSG-9 men opened the escape hatches and rear doors. One of the women terrorists, Hind Alameh, ran towards the rear door with a gun in her hand, but she was met with a burst of automatic fire from the first man through the door. She died instantly.

Disorientated and confused by the stun grenades, Wabil Harb staggered out of the cockpit and crashed into Suhaileh Sayeh as she ran for her life from the German soldiers chasing her down the aisle. Hit by a burst of rapid automatic fire, with some twelve bullets entering him, he too died instantly.

As he fell, one other group in the assault party broke through the starboard door into the aircraft just as Akache appeared from the cockpit. The first GSG-9 man opened fire with his sub-machine gun, cutting down the terrorist leader with eight or so bullets. But as Akache fell to the floor he dropped two hand grenades from which he had already removed the pins. Both grenades rolled into the first-class area and exploded. Scared and shocked, some hostages were screaming and others fainted. As the grenades exploded, the one surviving terrorist, Suhaileh Sayeh, opened fire with her handgun from the lavatory. One GSG-9 man returned the fire, hitting her in the chest.

Desperate to live, Sayeh screamed that she surrendered.

'Drop the gun, drop the gun,' the GSG-9 men shouted; then, 'Open the door, open the door,' as they heard the weapon drop to the floor. 'Kick the gun out, kick the gun out. Stand with your hands on your head.'

Outside the toilet two GSG-9 soldiers stood either side of the door, their sub-machine guns aimed at Sayeh's head from just three feet away. She came out shaking and pleading for mercy, and was ordered to lie on the floor. One man searched her for guns or grenades while two others stood over her, their weapons pointed down at her. 'If you move you will be shot,' she was told. The terrorist didn't move a muscle until someone came along, roughly pulled her to her feet and frogmarched her out of the door and down the steps to a waiting vehicle.

German doctors, nurses and psychologists, who had been brought out on the second Lufthansa flight, were on hand to take care of the deeply traumatised hostages. They realised the hostages had been forced to live through a most frightening experience during the one hundred and ten hours of captivity, terrorised by a gun-crazed killer, forced to sit in rapidly deteriorating conditions, forbidden to visit the toilet and all the time without sleep and given only a little food and drink. It would take weeks, months and, in some cases, years for the hostages to forget their ordeal.

For the combined Special Forces who took part, however, it was a brilliant and singular success. The dramatic rescue also proved to other European governments that, with training, discipline and courage, Special Forces could deal with terrorists and hijackers quickly and comprehensively. It also demonstrated to would-be terrorists that their operations might now well end in their own death.

Italy's Red Brigades came to prominence in December 1970 with the shocking and completely unexpected bombing of a Milanese bank, in which seventeen people were killed and fifty-eight injured. The Red Brigades were well organised, and consisted of cells of five people in Rome, Milan and four other Italian cities. There were fifty hard-core terrorists headed by a Strategic Directorate, supported by some five hundred unpaid part-time activists who rallied support for the left-wing terror group.

They targeted rich, prominent and influential Italians in various walks of life, assassinating a public prosecutor, a leading Turin lawyer, a judge and the editor of a national newspaper. They also succeeded in killing a senior officer of the Carabinieri in retaliation for the hard time that the military police force's counter-terrorism unit had been giving their members. A number of terrorists had been arrested and given long prison sentences.

But the Red Brigades weren't beaten yet. They retaliated with the brutal kidnapping of Aldo Moro, a former Italian Prime Minister and leader of the Christian Democratic Party, during daylight in Rome as he was being escorted by police officers. Moro's two-car convoy was ambushed in a narrow street and all three bodyguards and the two drivers were murdered in a hail of automatic fire. Moro was taken away. The kidnappers demanded the release of thirteen of the group's members, who were due to face trial, but the Italian government stood firm, refusing all the terrorists' demands. A massive police search was organised throughout Rome, but Moro was never found alive. Seven weeks later his body was discovered in the boot of a car. This act of terror caused revulsion against both the Red Brigades and most of the other terror organisations operating in Europe.

That single assassination made the Italian politicians sit up and take note, for they now feared for their own lives. They decided to get tough with the terrorists. The Carabinieri were permitted to set up a counter-terrorism organisation, the Nucelo Operativo Centrale di Sicurezza (NOCS), the members of which were nicknamed 'Leathernecks' because of the leather helmets they wore in action.

But further embarrassment would follow for the Italians. In December 1981 US Brigadier General James Dozier, serving with NATO headquarters in Verona, north-eastern Italy, answered his door at home and was immediately attacked and knocked unconscious by four men posing as workmen. The kidnappers drove him away in a truck and later announced they had captured General Dozier. The United States was absolutely furious and sent a six-man Delta Force team to Italy. The Carabinieri and NOCS were beside themselves, desperate to find the kidnapped Dozier, arresting any suspect they could lay their hands on. The anti-terrorist chiefs of the Carabinieri were certain that Dozier was still somewhere in northern Italy and that the terrorists were communicating with one another by wireless.

This was just the break needed, for the United States was able to provide the latest technical equipment to target the locating transmitters based in terrorists' safe houses. In a massive electronic surveillance operation, US helicopters criss-crossed the skies over Verona using these tracking devices. The United States also arranged for the orbit of its satellites to be altered so that they passed over northern Italy, and organised for the surveillance data to be downloaded as a matter of urgency.

Five weeks later the electronic surveillance operation pinpointed a block of flats in Padua. The Americans maintained constant electronic surveillance while the Italian counter-terrorism team kept watch on the ground. A woman employed by the Carabinieri's counter-espionage unit was sent to the flats, ostensibly selling household goods. Outside the targeted second-floor apartment she discovered a man standing guard and trying to look nonchalant. He told her that there was no one inside and that he was waiting for the owner to return.

It was agreed that the flat was the prison in which General Dozier was being held, and a ten-man assault team of Leathernecks was assembled at the back of the block, out of view of those holding him. All around the block, Italian sharpshooters kept watch, a US helicopter flew overhead and the Carabinieri arranged for two noisy bulldozers to pass the flats at midday.

As the bulldozers approached, the Leathernecks, armed with stun grenades, machine pistols and automatic assault rifles, crept up to the apartment. Two of them rushed the guard, silencing him with a blow to his windpipe and all but killing him. Two others, well-built and powerful, ran to the front door wielding sledgehammers and broke it down with four blows, allowing the squad to race into the flat. They hurled themselves at three terrorists who were in the hallway, smashing them to the floor, but one raced into a room where Dozier was chained to a bed, fumbling to grab a handgun. The terrorist was brought down by two officers, the handgun spinning out of his hand as they hurled themselves at him. He tried to make another grab for the gun, but was hit in the face. It was all over.

The dramatic rescue of General Dozier was rightly hailed as a major success for the Italian counter-terrorist squad and their methods. Dozier's life had been saved, and the terrorists had not only been caught alive and brought to trial, but they had also been persuaded to give information about other members of the terrorist organisation. It was the end of the Red Brigades.

It was fortunate that the Dutch government also took the decision to put together an elite counter-terrorist force – the BBE, Special Support Group – and they too took the SAS as their role model. It would not be long before this Special Forces unit was brought into operation.

In the late 1970s the South Moluccans were demanding independence for their homeland, the islands of the South Moluccas, in the Indonesian archipelago, which had been a Dutch colony before World War Two. After the war the Dutch handed the islands to Indonesia without any consultation about the wishes of the South Moluccans. Having no desire at all to be part of Indonesia, these people demanded their independence. They believed that only the Netherlands could bring about this change of nationality and so formed a pressure group to influence the Dutch parliament.

After five years of getting nowhere the South Moluccans decided they would need to take positive action to force the Dutch into helping them to achieve independence from Indonesia. As a result, in 1975 a militant group hijacked a train at Wijster, in the Netherlands, taking fifty-seven passengers hostage. Two passengers were killed by the hijackers as they stormed the train. The Dutch Army was detailed to surround the train and keep a permanent armed watch while the BBE were called in to plan and organise a rescue operation if it became necessary.

Two days later another South Moluccan unit of the same independence group stormed the Indonesian consulate in Amsterdam and held the eighteen staff hostage. The Dutch government and the BBE found themselves operating on two fronts, holding separate talks with the two groups while the BBE planned the best way to end both sieges. The train siege was called off after four days of non-stop negotiations, but the negotiations with the terrorists occupying the Indonesian consulate were far more difficult. The stand-off lasted two weeks and ended in a victory for the South Moluccans when the Dutch government agreed to start talks with their leaders.

But, as the talks dragged on, the South Moluccans realised that the Dutch appeared to be dragging their feet. And now, after more than two years of stop-start negotiations, they decided to add renewed pressure to the talks to show the government that their demands should be treated with more urgency. So, in May 1977, a group of nine South Moluccans hijacked a train between Assen and Groningen, in the northern Netherlands, holding fifty-one people hostage while another four terrorists occupied a school at nearby Bovensmilde, holding a further one hundred and ten hostages, nearly all of them children.

This time, the terrorists demanded not only independence from Indonesia but also the release of seven of their countrymen held in Dutch jails. They further demanded that an aircraft should be made ready at Schipol Airport, near Amsterdam, to fly the released prisoners and the two group of South Moluccans to an unknown destination. However, when the opening discussions took place, the hijackers felt the Dutch government negotiators were not taking their demands seriously, so they took the driver from the train, shot him in the head and threw his body on to the track.

This horrendous act shocked the Dutch government, media and people into a state of grave concern for the train passengers, but even more so for the school children, who, it was now clear, were at the mercy of killers. I was at the siege, watching the terrorists and the passengers on the train through binoculars and also watching the school buildings, but, for most of the time, the children were kept away from the windows and glimpses of the terrorists inside the school were few and far between.

Although Dutch soldiers were present in considerable numbers around the entire area, they were keeping a low profile so that they could not be seen by either the terrorists on the train or those holding the children. They did not want to frighten or provoke the terrorists into killing more hostages. For its part, the Dutch government was reluctant to send in the BBE, fearing that such a rescue mission could infuriate the terrorists and put both the children and the hostages on the train at even greater risk. As the days dragged on the mood of the nation changed. At first most people had given the government their full backing but now there was a widespread feeling that it should send in counter-terrorist troops to put an end to the sieges and, if necessary, kill the terrorists.

In the past these stand-offs with the South Moluccans had ended more or less peacefully, but the murder of the train driver had shocked the nation into the realisation that it was dealing with fanatics who were prepared to kill totally innocent people in their bid for independence.

There was one hope. The terrorists were permitting food and water to be sent into the school and the train. This gave the Dutch government and the professional negotiators reason to believe that there might be a peaceful outcome to the two sieges.

But the deadlock continued, and most of the people I spoke to demanded action because the wait had become unbearable. Even the mothers of some of the children taken hostage wanted the government to send in the BBE because they could not eat or sleep for fear of what might happen to their children. They spoke of the daily nightmare they were living, imagining their children terrified or, worse still, murdered by these terrorists, whom they hated with particular ferocity because they were prepared to put at risk the lives of innocent youngsters.

Debate about the situation raged throughout the country and across the civilised world as the days became weeks. The Dutch government was still fearful about the consequences for the children of using force. By contrast, those who would not have to take the blame if things went wrong were more keen to send in a rescue force. More and more people began to demand action, accusing the politicians of sitting on their hands rather than taking firm and decisive action.

Then steps were taken on two fronts. In the Netherlands, a train was brought into use so that the BBE unit could practise assault tactics and a school building was handed over to the men so that they could rehearse an attack. In Britain, the SAS was put on standby in case the Dutch government asked for their advice, and, if a rescue operation were to go ahead, their active assistance. At one time the SAS men were moved to an airbase in Britain, in readiness to fly in and lead an assault on both the school and the train.

And then, totally unexpectedly, came the breakthrough. The terrorists inside the school could no longer take the tears and the pleas for freedom of the young hostages. They agreed to release all the children and hold just four teachers. The scenes I witnessed brought tears to the eyes of everyone, including the police and the soldiers. Most of the children came out walking, desperately searching the faces for their parents, some crying, many in a state of shock, some hardly able to walk while others ran into their parents' arms. The anxiety and pain on the faces of the waiting mothers turned to joy and smiles, mixed with uncontrollable tears, as they threw their arms around their children, hugging and kissing them.

After all the children had left the school without injury, the plight of the train hostages took centre stage. The four teachers were still being held, but it seemed that the terrorists at the school were now in a more peaceful frame of mind. However, reports from those permitted to take food and drink into the train spoke of mounting anger and frustration among the terrorists, some of whom appeared to be growing increasingly hysterical and violent at the lack of response to their demands. The negotiators told the government forcefully that the time had come to make a decision – to either accept some of the terrorists' demands or order an attack on the train to lift the siege. If no decision were taken, the negotiators believed, the terrorists would start killing the hostages.

Finally, after days of prevarication, the Dutch government gave permission for the BBE to launch simultaneous assaults on both the school and the train.

Before dawn on June 11 Royal Netherlands Air Force F-104 Starfighters swooped low over the train, using the noise of their afterburners to frighten and confuse the terrorists into thinking the war planes were about to attack the train with rockets. As the jets flew a number of sorties along the length of the train the BBE teams moved to either side of it, unseen by the terrorists and hostages on board. At a given time a green flare was fired into the sky, the signal for the assault teams to detonate the charges they had placed against a number of the train doors. As the charges detonated in a series of crashing explosions which could be heard a mile away, the assault teams, armed with sub-machine guns and handguns, burst into the carriages.

Shaken and unnerved by the dramatic explosions, some of the hostages screamed in terror, believing the train had been blown up. Two terrorists raced towards the first soldier who entered the train, but as they were drawing their handguns he opened fire, killing both of them in a hail of bullets.

Other soldiers were yelling at the hostages to get down and take cover. Two more terrorists raced out of the first carriage to come under attack and ran into BBE marksmen running towards them inside the train. The terrorists fired two shots before they too were gunned down. Three others took cover behind two rows of seats and exchanged fire with the BBE men, who were still further down the same carriage. In between the terrorists and the soldiers, some eight hostages were lying flat on the seats in the hope that the backs of these would offer protection from the bullets ricocheting around inside the train. Some hostages were screaming in terror, others shouting for help, while the BBE team kept yelling at the hostages to stay still and keep down.

By now all the windows of the carriage were shattered, and to the hostages it felt like pandemonium. And then two of them scrambled to their feet in a bid to escape the carnage. Believing they were terrorists bent on attacking them, the soldiers shot both of them dead.

The three terrorists pinned down at the end of the carriage continued firing at the soldiers, reloading and then letting off another magazine. A decision was taken to attack the trio from the other end of the carriage and another BBE team was sent in to do so. A signal was given to the first team to stop firing the instant the second team crashed their way through the door into the carriage. The terrorists realised what was happening too late and, although they managed to fire off a couple of rounds, they were met by a non-stop rain of bullets from the first two soldiers to burst through the door.

It was the end.

At the same time as the green flare had lit up the dawn sky another BBE unit in two armoured personnel carriers had raced through the gates leading to the school, then through the grounds, coming to a halt by the main entrance. The soldiers jumped from the APCs, smashed through the main door and raced to the classroom where they knew the terrorists and hostages would be found.

They burst into the room, yelling at everyone to lie down. Within seconds eight BBE men were standing in the room, their guns pointing at the eight men there – four hostages and four South Moluccans. All immediately obeyed, and as some of the BBE men kept their sub-machine guns trained on all eight of them, others moved from man to man, checking their clothes for weapons and grenades. One by one the eight men were ordered to their feet and searched again, and then the hostages were separated and handed over to medics and support staff, who took them away to hospital and a meeting with their families. The South Moluccans were handed over to the police and escorted away under armed guard. Not a shot had been fired by the terrorists from the moment the BBE's assault had begun.

The Soviet Union was not slow in following western European countries' creation of Special Forces and, to a certain degree, it too was influenced in this by the success of the SAS. In fact the use of special soldiers was not new to the Russians, for during World War Two the Red Army had developed the Spetsnaz, the Spetsialnoje Naznachenie, or forces of 'Special Designation', who had no real equivalent at that time. The original purpose of the Spetsnaz was described as 'diversionary reconnaissance', as these units were sent ahead of the main assault force to sabotage the enemy's installations, defences, ammunition dumps and vital communications.

During the Cold War the Spetsnaz brigades were trained for even more adventurous missions, such as sabotaging NATO nuclear weapons sites, laying mines in enemy territory, always going ahead of the main army to cause disruption and damage.

After the Soviet Union invaded Afghanistan in 1979, the Spetsnaz carried out many such operations in that notoriously inhospitable terrain.

Since Afghanistan, Spetsnaz forces have become a well-established and more open Special Forces unit. Previously they had always worn the uniform and insignia of the Soviet airborne forces, but this has now changed. Their uniform now reveals that they are members of the elite Spetsnaz, and those who join the unit are put through extremely tough training schedules and treated with great respect by the Russian defence chiefs.

Since the creation of the Russian Confederation of Independent States, most Spetsnaz forces have been withdrawn to within Russia's borders, while many of the former republics have created their own, smaller Special Forces units. Today the Spetsnaz comprises three main elements. There is a brigade-sized formation responsible for reconnaissance which operates in battalion or company strength; a brigade-sized formation which operates in small, eleven-man teams similar to the SAS; and a single battalion divided into two companies, one for long-range reconnaissance and the other for airborne operations.

The rebellion of Chechnya against the new Russian Federation provided the Spetsnaz with what it saw as an excellent opportunity to show its full expertise and professionalism and to put into practice all its training, which it considers to be among the toughest of any Special Forces in the world.

In 1991 the fiercely independent people of Chechnya elected Dzhokar Dudaev, a former Soviet Air Force commander, as president of the former Soviet republic. The Muslim Chechens rallied around their new hero-president, who was determined to make the nation a fully independent state. The Russian President, Boris Yeltsin, was equally determined that Chechnya should remain within the Russian Federation and in December 1994 deployed troops and armour against President Dudaev.

But Yeltsin found the Chechens a tough breed and, despite the Russians' military superiority in tanks, artillery, aircraft, helicopter gunships and personnel, the Chechens refused to lie down and accept defeat. Within twelve months Grozny, the capital of Chechnya, had been reduced almost to ruins, bombed from the air and shelled from the ground, but still the Chechens held on. They took to the hills and the mountains and began a blistering guerrilla war against the Russian forces on the ground, attacking their camps by night and fading away in the dawn.

Into this chaos the Spetsnaz forces were frequently sent on special operations. Sometimes they were parachuted into Chechen-held territory in the hills outside Grozny; at other times they went into the city itself on a mission to grab Chechen fighters and retreat to their own lines with their captives.

In one such raid in 1995 two Spetsnaz companies each of about thirty men were tasked with attacking a former government office block on the outskirts of Grozny which was being used as a headquarters by Chechen rebels. The objective was to storm the building, kill all those inside, set fire to the place and then retreat. One company was tasked with controlling and holding the immediate area surrounding the block while the second company stormed it to clear it of rebel fighters before setting fire to the building.

It was a cold autumn dawn when the Spetsnaz operation began. Three Russian Army T-72 tanks led the way through the deserted streets to the target building, and behind them some twelve APCs, armed with single heavy machine guns, carried the Spetsnaz assault force. As the convoy arrived at the square where the Chechen headquarters stood the tanks opened fire on the building while the Special Forces soldiers disembarked from the APCs and took over the square and the streets leading off it. They met virtually no resistance except for the occasional sniper round.

After five minutes of intense fire from the tanks the assault force stormed the building. The sound of automatic fire could be heard from those keeping watch outside, as well as frequent explosions as the Spetsnaz troops tossed grenades into the rooms before moving inside. Those outside could also hear the odd scream and, sometimes, orders being barked out by those commanding the assault team. The basement rooms were cleared first with a dozen or so hand grenades, CS gas and smoke grenades. As Chechen fighters emerged, coughing and spluttering their way up the stairs, the Spetsnaz showed no mercy. Everyone emerging through the smoke and CS gas was shot as they came into view. The Spetsnaz were under orders to take no prisoners.

The Special Forces troops moved from floor to floor, but the Chechen fighters were determined to hold out as long as possible. As the Spetsnaz reached each floor the Chechens would throw grenades down on to their heads and spray them with automatic fire. But these attackers were not the usual underfed, under-trained and underpaid Russian conscripts who only wanted to get back home safely but the Federation's crack troops, privileged, well paid, well trained and keen.

When the Chechens reached the top of the eight-storey building they barricaded themselves into a large room and threw cabinets and shelving on to the floor to offer some protection against the attackers. Then they waited.

The assault on the top-floor redoubt began with an explosive charge which disintegrated the wooden doors, followed by half a dozen grenades and a non-stop torrent of bullets from machine guns which sprayed the room for several minutes. The Chechens fired back and killed at least three Spetsnaz troops during this final attack but, in reality, they stood no chance. It is not known whether any of the Chechens surrendered or tried to surrender, for all were killed, ruthlessly and professionally. And some of the bodies were thrown out of the top windows, to the cheers of the

Spetsnaz men keeping guard in the square below.

Yet, for all its overwhelming strength, the Russian Army was forced to pull out of Chechnya because of the determination and courage of the Muslim fighters in their guerrilla attacks on the dejected, down-hearted ordinary Russian soldiers, who had no stomach for the war. Despite horrendous losses, a devastated capital city and many war-ravaged towns and villages, the small army of Chechen fighters, supported by a half-starved local Muslim population, had been able to absorb the worst their enemy could throw at them and still inflict so much damage on the Russian forces that they decided first to retreat and eventually to abandon their campaign in Chechnya.

CHAPTER 5

RESCUED

HOSTAGE RESCUE OPERATIONS are a key element of Special Forces work, and one of the most impressive to date took place in 1976, when an Israeli elite squad stormed a hijacked Air France airbus at Entebbe, in Uganda. That raid spawned books, films, television programmes and lectures and has become part of the folklore of the world's Special Forces.

The hijacking of Air France Flight 139 was meticulously planned by leading members of a number of terrorist groups who, for this single dramatic event, decided to work together in an attempt to force the French and German governments to free terrorists they had jailed. The prisoners included members of the West German Baader-Meinhof gang, the Popular Front for the Liberation of Palestine, Black September (which grew out of the PFLP) and the Japanese Red Army.

Behind the hijacking was a West German lawyer, Wilfried Boese, a fair-haired, blue-eyed, intelligent, sophisticated man of twenty-eight who had previously worked for the PFLP with his friend the terrorist Carlos Ramirez, better known as Carlos the Jackal. For this mission he recruited twenty-four-year-old Gabrielle Kröcher-Tiedemann, a stocky, dark-haired, aggressive West German woman who had joined the Baader-Meinhof Gang as a teenager. She had taken part in the attack on the OPEC ministers in Vienna in December of the previous year. The other two members of the team were Arabs, tried-and-tested PFLP gunmen. Boese turned to a South American friend of Carlos, Antonio Degas Bouvier, who had planned the raid at the Munich Olympics, to mastermind the hijacking.

On the morning of June 27 Boese and Kröcher-Tiedemann drove from an apartment in Kuwait and bought first-class tickets to Paris via Athens for a flight due to arrive at Athens at 7 am. The two Arab gunmen arrived at the airport separately and bought tourist-class tickets for the same flight. They were each carrying a tin of dates containing a 7.65mm Czech automatic pistol and a grenade.

On arrival at Athens they waited for the Air France flight to Paris which had begun its journey in Tel Aviv. On board were two hundred and fifty-eight passengers, including about a hundred Israelis, and a crew of twelve. In the toilets at Athens one of the Arabs passed Boese the pistols and the grenades and, undetected, the four boarded the plane.

The aircraft had been in the air only some ten minutes when Kröcher-Tiedemann rose to her feet, stepped into the centre aisle, shouted for attention and raised her arms high above her head. In each hand was a grenade. 'Sit down,' she yelled. 'Everyone must sit down.'

At that moment the well-dressed Boese walked up the aisle waving a pistol at the stewardess and entered the flight deck. The two Arabs, one wearing a red shirt, the other a yellow one, rose from the seats and ran down the aisle, each brandishing a pistol. There were screams of terror from the passengers, who then heard a voice announce over the intercom, 'We are Palestinians. If you remain seated and do as you are told no one will be harmed.'

The two Arabs moved to the exits, carrying what looked like

boxes of chocolates, and fastened them to the doors. The voice on the intercom – it was Boese's – announced, 'The boxes contain explosives. If there is any trouble these will be detonated.' The German then lectured the passengers, telling them the reasons behind the hijacking and explaining that they would be held hostage until a number of freedom fighters had been released from jails in Israel and Europe. He emphasised that the hijackers bore no animosity towards the hostages, who would be in no danger as long as they obeyed orders. The hostages listened to the speech in silence.

An hour later the Air France plane landed at Benghazi, in Libya, refuelled and took off for Uganda. Some two hours later the big jet flew low over beautiful Lake Victoria and landed at Entebbe. Believing their ordeal was now over, the passengers cheered and clapped and soft drinks were brought to the plane. The stewardesses handed them out to everyone as though nothing untoward had occurred. The passengers began to relax and chat among themselves.

Some two hours later they were taken off the plane and made their way under a guard of Ugandan Army soldiers to the rather dilapidated, two-storey terminal building. They were welcomed by an airport official who offered them drinks and sandwiches, promising them a meal later. Within an hour or so President Idi Amin, dressed in combat fatigues, arrived by helicopter, accompanied by his four-year-old son, who was dressed likewise. Amin introduced himself, explaining, 'For those who don't know me I am Field Marshal Dr Idi Amin Dada, President of the Republic of Uganda.' He had come a long way from his days as a sergeant in the British Army in Uganda. Laughing and smiling, he mingled with the hostages. He shook hands with some of them, saying, 'You must not worry. I will take care of you like a father. I will see that you are all released safely.'

But the hours of waiting became days and still there seemed to be no progress in the negotiations which the hostages understood were going on between Uganda, France, Britain, Greece and Israel – citizens of all of which countries had been on the flight – and Antonio Degas Bouvier, the mastermind who had flown to Kampala to conduct the negotiations on behalf of the hijackers. Progress was slow because all negotiations were being conducted through Somalia and then Idi Amin, who insisted on the role of mediator, before Bouvier was consulted about his next reply.

On the third day the Israeli Cabinet agreed to release a few Palestinian prisoners in return for the safe return of some of the hostages. In fact they were playing for time. Bouvier agreed and the hijackers set free most of the women and children hostages and a few elderly men, some one hundred and fifty people. These were immediately flown to Paris but there was not a single Jew among them. Every single hostage now held at Entebbe was Jewish. As a result, the Israeli Cabinet came to the conclusion that the only way they could save the lives of the remaining hostages was by direct military action.

The man chosen to command the force was Israel's senior operational officer, Brigadier Dan Shomron, who had been a paratrooper during the Six Day War and a tank commander during the 1973 war. Every Israeli officer involved in the detailed planning of the rescue, as well as those who would take part, knew that unless they could be in action around the Entebbe terminal within one minute of landing, the lives of the hostages would be in grave danger.

The assault force was led by Lieutenant Colonel Yoni Netanyahu, an American-born Israeli paratrooper with a glittering military record. In the 1967 war Netanyahu had commanded a mortar platoon on the Syrian front and in fierce fighting was hit twice in either arm. But he had refused to pull back and somehow continued to fire.

Netanyahu had learnt all about the man who, in the 1930s,

founded the famous Israeli clandestine Special Force originally known as the Jewish Night Squads and so called because they attacked at night, returning to base before dawn. That man was a famous British soldier and guerrilla leader, Orde Wingate. A religious mystic, Wingate trained at Sandhurst and in the 1920s served with the Sudan Defence Force, becoming intoxicated with desert life. In the 1930s he moved to Palestine and Transjordan, in both of which he trained Jewish and Arab special soldiers. Wingate won fame in Burma in World War Two when he founded the famous Chindits, a jungle-trained Special Force who caused great problems for the Japanese by fighting guerrilla style behind their lines.

In addition to learning about the methods of the Jewish Night Squads, Netanyahu had also been one of thousands of young Israeli soldiers who had rushed to the Golan Heights to face the Syrian tanks when the 1973 war broke out. Having helped to throw back the Syrian Army, he and his fifty hand-picked paratroopers were then detailed to back the Israeli tank thrust deep into Syrian territory. At night on the Golan Heights he would take out a small squad of paratroopers to pinpoint enemy tank concentrations.

Netanyahu and his men believed that the greatest risk of the entire Entebbe raid – code-named Operation Thunderball – would be the two thousand-mile flight from Tel Aviv to Entebbe. They feared their slow-moving Hercules C-130 transport planes might be intercepted by Saudi Arabian or Egyptian fighter jets and ordered to land. Six Phantom jets escorted the Hercules as far as their fuel tanks would permit, the pilots all the while searching the night skies for attack fighters. But there were none.

Three Hercules C-130s were taken on the mission in case one was shot up on the tarmac. Both the hostages and the Israeli Special Forces troops would then be at the mercy of the militant hijackers, the Ugandan Army and their leader, the mercurial

Idi Amin. In one of the Hercules was a black Model 60 Mercedes 220 – precisely the same model and colour as Idi Amin's personal car – which the Israelis hoped might fool the Ugandan soldiers guarding the airport at Entebbe. This Mercedes had been found rusting in a Tel Aviv junkyard and in forty-eight hours had been sprayed black, polished, totally serviced and given new tyres and a new leather interior. A fourth Hercules – an Israeli mobile hospital equipped with surgeons, doctors and nurses – landed at Nairobi.

And ahead of the three Hercules was a Boeing 707 aerial command post – crammed with sophisticated electronics – which would take up its position above Entebbe. In command of the entire rescue mission at some thirty thousand feet was General Yekuti Adam.

At one minute past midnight on Sunday July 3 1976 the pilot of the first Israeli Hercules called the control tower at Entebbe, saying, 'This is El Al Flight 166 with the prisoners from Tel Aviv. Can I have permission to land?'

However, the two Ugandan flight controllers had not seen on their radar screens two more aircraft skimming the surface of Lake Victoria and making for Entebbe. The first Hercules touched down on the runway and headed towards the terminal. It came to a halt and the heavy ramp of the cargo hold slowly sank to the ground.

'Go,' yelled the commander.

A Land Rover was first down the ramp, roaring at speed on to the well-lit area near the control tower, followed by the black Mercedes. This was followed by a second Land Rover. As the three vehicles swept steadily the four hundred or so yards to the terminal, where the hostages were being held, the Ugandan guards on duty snapped to attention.

As soon as the small convoy came to a halt and Israeli paratroopers with Uzi sub-machine guns clambered out of the vehicles, one Ugandan soldier shouted an alarm and raised his rifle. Before he could take aim he had been shot through the head.

As the first shot was fired Netanyahu and his fifty paratroopers ran down the ramp of the Hercules and fanned out across the area. As they took up positions and began to move more slowly towards the terminal they could see an intense firefight surrounding the Mercedes and the two Land Rovers. Yoni and seven other armed commandos in the Mercedes were raking the Ugandan troops with their automatic assault rifles, while the driver of one of the Land Rovers – a sniper – was methodically shooting out the lights along the roof of the terminal.

Within less than a minute the Ugandans turned and ran in fear of their lives, leaving a dozen of their number lying wounded or dead on the ground. But they ran into Israeli paratroopers who were approaching the terminal from the other side. Another firefight began, with the Ugandans not knowing where to flee to escape the terrifying onslaught.

Within seconds Kröcher-Tiedemann and Boese were dead – Kröcher-Tiedemann gunned down as she raised her pistol; Boese cut down as he raced out of the building to see what was going on.

Now there were eight more hijackers to be accounted for. Netanyahu and two paratroopers raced along the corridor and up the stairs to the first floor. Before them they saw a door and heard gunfire behind it. Netanyahu took up a position next to the door while opposite him the two paratroopers waited, grenades in their hands. He kicked open the door and jumped back as his two comrades hurled the grenades into the room and a burst of machine-gun fire came from inside. As the grenades exploded, Netanyahu and his men ran into the room and raked it with their Uzis. The two hijackers never stood a chance.

Down on the ground two Palestinians were seen running from

the terminal. They covered only a few yards before both were killed in murderous fire.

Inside the building the Israeli paratroopers were shouting in Hebrew, 'Lie down, lie down, stay still.' In panic, two hostages jumped to their feet and the Israelis, believing them to be hijackers, shot them dead.

Meanwhile Netanyahu and the two paratroopers were racing towards the terminal's roof terrace, having been briefed that a detachment of Ugandan soldiers usually kept guard there. There they found a few Ugandan soldiers, who within seconds were dead. Other Israeli paratroopers followed Netanyahu and his men up the stairs and during the next five minutes firefights broke out in corridors and on stairs.

Suddenly the horrendous noise that had reverberated around the building ceased and there was silence. Netanyahu checked his men, all of whom were safe and well except for three with minor gunshot wounds. He talked to General Adam in the aerial-command Boeing above the airfield, telling him, 'The terminal is secure. We are now going to evacuate the hostages.' But he had barely finished speaking when a single shot rang out and he was hit in the back. He was dead before he touched the ground.

Already shocked that their Commanding Officer had been killed, the Israelis were surprised again seconds later when a new barrage of fire could be heard coming from the other side of the airfield. The Israelis had already placed paratroopers around their aircraft, fearing a counter-attack. They knew only too well that if the Hercules did not get off the ground with the hostages aboard they would be shown no mercy. Not one of them would survive.

There was one further worry that the Israelis would have to attend to. Mossad, Israel's famous secret service, had reported to Tel Aviv that sixteen Russian MiG fighters were on the ground at Entebbe. It was understood that these were in good condition and flew frequent practice missions. They would need to be destroyed before the Israelis flew out of Entebbe, otherwise the lumbering Hercules C-130s would be at their mercy once in the air.

From the cargo bay of one of the other Hercules an APC and four jeeps came roaring on to the tarmac. On the bonnet of each jeep was a heavy machine gun which raked the area. Whenever they saw Ugandan troops or any vehicle moving around, the machine gunners opened fire with blistering attacks. Following these vehicles were one hundred Israeli commandos armed with Uzis, who raced towards the terminal and took up positions supporting those who had gone in on the first attack.

Another unit, armed with explosives, was dispatched to the section of the airfield where the two squadrons of MiGs were parked. Their task: to wreck the fighters.

But the Ugandans were not finished yet. Although many had fled, there were others, better trained, who had decided to stand and fight. Rapid fire was now raining down on the Israelis from Ugandans who had made their way to the roof of the control tower. The Israelis returned fire, but there was little chance of their hitting their targets high above them. The armoured car replied to the enemy on the roof with devastating bursts of fire and an RPG hit the top of the control tower with a flash and a blinding explosion. Paratroopers now raced towards the terminal. They knew they had to stop the Ugandans firing before any hostages could be rescued.

Commandos raced up the stairs armed with grenades. The first Israeli was shot through the chest, but two of his mates hurled grenades through the open door of the control room and waited. Seconds later the grenades exploded, ending all resistance, and the Ugandans surrendered. As they left the tower the Israelis threw four more grenades into the room, slammed the door and fled down the stairs.

The commandos ordered to destroy the MiGs were having a

tough time. As they approached the fighters they encountered heavy automatic fire from the Ugandan guards. They replied with grenades which caused fires to burn around the MiGs, lighting up the immediate area. The Ugandans withdrew into the dark and continued firing at the Israelis, keeping them away from the aircraft. After some thirty minutes the Israelis had succeeded in setting the MiGs alight and they continually peppered them with gunfire. When they pulled back they knew there was no danger that the fighters would be able to follow the Hercules into the sky.

The Israelis had now been on the ground some forty minutes and they hadn't even yet started to move the hostages out of the terminal, where they were lying in fear on the concrete floor with a dozen paratroopers standing guard over them. The hostages could still hear sporadic gunfire going on all around the airfield, which meant the Ugandan forces were not yet defeated.

A problem had occurred with the refuelling of the Hercules and, as the minutes ticked by, the decision was taken to abandon this procedure at Entebbe and head for Nairobi. The commanders were advised that the transport planes should have sufficient fuel left in their tanks to fly to Nairobi but no further. In any case, staying on the ground in the hope of fixing the refuelling equipment was probably a greater risk. No one knew what forces Idi Amin might have in reserve to throw into the battle, and if the Israeli aircraft were hit on the ground and put out of action there was the possibility of a terrifying massacre.

Teams of Israeli troops took off in their jeeps and Land Rovers to check the entire perimeter of the airfield for any Ugandan troops gathering for a counter-attack. They found none. It seemed the troops that hadn't been killed or wounded in the firefights had fled into the surrounding bush. As soon as the Israelis returned the order was given to bring out the hostages from the terminal building and take them to the planes. A hundred paratroopers formed a protective guard for the frightened hostages, some carrying children, as they stumbled and ran from the terminal to the nearest Hercules.

As the plane prepared to take off Israeli commandos in their jeeps took up positions at the side and the end of the runway, checking that no Ugandan soldiers were preparing to take shots at it as it roared down the runaway. Other soldiers checked every room in the terminal, to make sure there were no hostages left. A final count was made on board the plane before take-off. Exactly fifty-three minutes after landing at Entebbe the first Hercules, with all the hostages on board, took off. The last Hercules, carrying the Israeli rearguard and the black Mercedes, lifted into the sky one hour and thirty-five minutes after landing.

Four years earlier, in 1972, the world had been shaken by another hostage crisis when eight Black September terrorists managed to penetrate security and take over the Israeli team's living quarters at the Olympic Games in Munich. Two Israeli competitors, the weightlifter Yossef Romano and the wrestler Moshe Weinberger, were shot dead and eleven more members of the Israeli squad were taken hostage. The world was able to follow every move of the terrorists and the West German police in live television coverage as the drama unfolded. Black September demanded the release of two hundred Palestinians held in Israeli jails, but the Israelis refused point-blank to consider the demand. The embarrassed West German government agreed to give the terrorists, accompanied by the hostages, safe passage to Egypt and flew them all in two army helicopters to Fürstenfeldbrück military airfield.

But the West Germans had laid a trap. Police marksmen were lying in wait at the airfield and, as the terrorists and their hostages disembarked, police snipers opened fire with rifles, intending to take out the eight terrorists in one hit. In fact five terrorists were hit, two of them killed outright and the three others merely wounded. The remaining three terrorists weren't finished yet. In the chaos surrounding the helicopters the three terrorists and their wounded colleagues dragged the hostages back to the helicopters as police armoured cars raced across the tarmac in a bid to prevent the terrorists forcing the hostages back on board. When the armoured cars were twenty metres away the terrorists turned their sub-machine guns on the hostages inside the choppers.

As the armoured cars screeched to a halt and armed police officers leapt out, the two helicopters exploded and burst into flames, killing everyone on board. At the final count all the hostages, five terrorists and one policeman were killed. Three terrorists survived and were captured. This attempt by police to execute a Special Forces-type hostage rescue had ended in total disaster because every single hostage had been killed. Governments around the world learnt a vital lesson that day. National police forces were simply not trained to tackle this type of situation. If governments were to successfully combat and defeat terrorism in actions where hostages had been taken, it was necessary to train elite counter-insurgency forces to an exceptionally high standard. These forces would lead the fight against international terrorism efficiently and ruthlessly. From that day on it was accepted that Special Forces within the army would be responsible for such missions.

Mossad is principally Israel's foreign intelligence-gathering organisation, based originally on MI6, Britain's Secret Intelligence Service. However, it also has a fearsome Special Forces arm which has been involved from time to time in organising assassination, kidnapping and bombing of terrorists. It has also taken part in raids on suspected terrorists' homes, offices and headquarters, usually employing the same tactics, precision and professionalism as the SAS on comparable missions. Mossad was ordered by the Israeli government to hunt down and kill those Black September terrorists who organised the massacre at Munich. Instructions were given that no matter how long it took to track down and kill those responsible, the athletes who died at Munich would be avenged.

One month after Munich the first Black September terrorist was dead. Wael Zwaiter, Al Fatah's official in Italy, was shot by two members of a Mossad hit squad of four men in the lift of the block of flats where he lived in Rome. That day Zwaiter arrived at his home alone and walked into the lift. Before the gates shut, two men stepped from the shadows and blasted him at point-blank range with handguns. The Mossad agents walked to a waiting car. There were no witnesses.

Some weeks later Mossad executed the second man, Mahmoud Hamshari, the Palestine Liberation Organisation's senior official in Paris. This killing was more ingenious. One night the Mossad hit men gained access to Hamshari's office and set up a remote-controlled telephone bomb. The following day Hamshari was working at his desk in his well-guarded office when he answered a phone call. As he picked up the receiver the bomb exploded, killing him instantly. Mossad had not completed its task yet, but by now Black September representatives across Europe were very worried.

In early 1973 Dr Bassel Rauf Kubeisy, a leading organiser of the Popular Front for the Liberation of Palestine in Paris, was tracked down and shot dead by a Mossad agent as he arrived home one dark night. He was shot in the back with a single bullet. The Mossad agent melted away into the night. And again there was no witness to the execution.

A few months later the man whom Mossad believed had masterminded the Munich massacre, an Algerian named Mohammed Boudia, was blown to pieces as an explosion ripped apart his car, which he had parked near the centre of Paris. One month later Ali Hassan Salameh, who had been responsible for a number of Black September operations in Europe, was tracked down by Mossad in the small market town of Lillehammer, in Norway.

The Israelis believed that Salameh was one of Black September's most senior planners, responsible for directing and organising many successful missions. For this operation Mossad sent a team of eight agents to Oslo. They were taking no chances. They spent two weeks checking out the intelligence they had been given, masquerading as businessmen visiting Norway to purchase newsprint. Some of the group travelled up country while others stayed in a five-star hotel in the capital. They had been given the exact location of the wanted man, who lived a quiet life in Lillehammer, but they did not have the precise address.

Discreet enquiries were made and the Mossad agents discovered that Salameh lived with a Norwegian woman in a block of flats in a middle-class suburb of the town. They checked out the location and for two days kept watch on the apartment. They saw a swarthy man of North African appearance leaving his home in the morning and returning each evening. They confirmed that he was living with a Norwegian woman.

After two weeks the two Mossad gunmen were given the orders to take out Salameh. They decided to assassinate him by shooting him in the back as he returned home one evening in July 1973. Their escape was carefully planned because Mossad wanted to make sure their hit men were out of the country as quickly as possible so that the Norwegians would be unable to arrest them and put them on trial.

The Israeli government, which was kept apprised of every Mossad attack before it took place, knew that the Norwegian government would not take kindly to Mossad agents assassinating a man in cold blood, even though he was a member of Black September. Norway would have required the arrest of the suspect and then for Israel to seek leave to have him deported to Israel for his alleged offences through the due process of the Norwegian courts. But the Israeli government had no wish to obey the laws of Norway, or any other country for that matter. They wanted vengeance for the murderous attack on their Olympic team and they were determined to kill as many Black September suspects as possible.

On the evening of July 21 the Mossad gunmen were taken by car to the road where Salameh lived. They waited inside the car for his return, knowing that he arrived home at around 7 pm. Punctual as ever, the North African man was dropped off outside the block of flats by a friend, who then drove away. Waiting for him in the flat was his wife, who watched him get out of the car and walk towards the building.

She saw two men walk up behind her husband. One of them took out a gun from his shoulder holster and shot him in the back at point-blank range. She watched as her husband collapsed in a heap on the ground and the two men walked away. She immediately called the police and the ambulance service, then raced out to see her husband. But he was dead. Seconds later the Mossad gunmen were picked up and driven at speed the one hundred and fifty miles to Oslo, where they caught the first plane out of the country.

But Mossad had made a mistake. The man it murdered was not the Algerian Black September activist Ali Hassan Salameh but a totally innocent Moroccan named Ahmed Bouchiki, who was married to the Norwegian woman. To the world of secret services the Mossad blunder was a disgrace. It seemed extraordinary that Mossad's eight agents tasked with the assassination never bothered to check the identity of Ahmed Bouchiki. It was such a rudimentary error. Mossad, the secret service that prided itself on ruthless efficiency and impeccable intelligence, had made a fundamental error that had not only killed an innocent man but also delivered a massive blow to its own reputation.

The Norwegian police acted with great speed. They deployed armed police around Oslo international airport and sent another group of armed police to the hotel in the city, where they knew a number of 'Israeli businessmen' had been staying. Though the two gunmen had escaped the net, the other six were arrested. Despite representations from the Israeli government at the highest level, the Norwegian government refused to bow to pressure and all six men were put on trial, convicted and sentenced to between two and five years in jail. All apologised in court for their appalling error. In fact none of the six Israelis served more than two years of their sentences.

Stunned by the blunder of its Mossad agents, the Israeli government put a halt to the plan to wipe out every Black September activist believed to have had a hand, even in the planning, of the Munich operation. The United States and several European governments, aware of the covert programme of assassinations, told the Israeli government that it should cease. Through normal diplomatic channels the European governments made it plain that they could no longer condone any killing of people in their countries.

Israel agreed, but secretly continued to track down Black September activists. Some six years later Mossad discovered that Salameh was living in Lebanon, and agents posing as Lebanese tracked him for two months. His apartment, his routine, his colleagues and the restaurants where he ate and the people he met were all discovered. In January 1979, as Salameh was driving his Mercedes towards Beirut from a village in the hills that he often frequented for meetings, Mossad set off a massive bomb parked in a car by the side of the road. At the precise moment Salameh drove by, the car exploded, destroying the two vehicles. Salameh was blown to pieces. It had taken a long time but Mossad had finally got its man.

CHAPTER 6

HEROES OF THE SEA

THE MEN MAINLY RESPONSIBLE for Britain's success in the Falklands War of 1982 were the Royal Marine Commandos of the Special Boat Service, who provided the vital intelligence necessary for the successful invasion of the island. The risks they took have never been fully recognised, for those SBS men were landed on the Falkland Islands before war had been declared and therefore, if any had been captured, the Argentines would have been within their legal rights under the Geneva Convention to treat them as spies rather than prisoners of war. And the British government would have had little or no legal recourse to prevent their execution if they had been found guilty of spying.

But every member of the SBS who landed on the Falklands knew the great risk they were taking in this respect and realised that each of their many missions behind enemy lines was fraught with danger. To add to the perils facing them, they were working almost entirely alone, with no back-up.

The Falklands conflict was tailor-made for the tactics and techniques of the SBS, which was in its element and rose brilliantly and courageously to the challenge. Although its vital task in the Falklands was reconnaissance, it was also responsible for carrying out other typical Special Forces roles, including raiding, initial assault and deep penetration.

But the SBS's first involvement in the Falklands War was a defensive role against overwhelming odds. A tiny Royal Marine Commando unit of some twenty-two men was guarding South Georgia, the largest of a small group of islands on the edge of Antarctica. On the morning of April 1 1982 they awoke to find that an Argentine frigate, the *Guerrico*, had entered Grytviken harbour on South Georgia and was ordering them to surrender, otherwise it would open fire on them with their powerful guns.

In command of the British force was Lieutenant Keith Mills , who rejected the order to surrender. Despite the fact that his men were facing a David and Goliath conflict, the young Royal Marine Commando ordered them to open fire against the frigate with their rifles, light machine guns and 66mm rockets. Of course, these weapons were useless against a frigate with its powerful guns, but the *Guerrico*'s captain was uncertain exactly what soldiers and artillery the British force had at its disposal.

But worse was to follow. As the Argentine ship steamed out of the harbour the Marines rapidly constructed what defences they could, fearing a full artillery barrage. Lieutenant Mills opened up his wireless connection with the British forces on the Falklands, informing them of what had happened. He also told the senior British Army commander that they had repulsed the first Argentine attack but were expecting renewed action. They didn't have to wait long.

As soon as the *Guerrico* had withdrawn out of range of the light British weapons, it opened up with its powerful 100mm guns. Though the British defences were taking a ferocious battering, the Marines refused to surrender. Mills believed that if the Argentines were to launch an attack and land forces from the ship, his men would be quite capable of putting up serious resistance. He guessed that the troops on board the frigate would not be battle-hardened and hoped that they would not include any Special Forces.

Mills discussed the desperate situation with his men, giving his personal opinion that they couldn't hope to hold out without further support from either Royal Navy warships or British warplanes. He explained that there were no RAF warplanes on the Falklands to come to their rescue and so there was little chance of their withstanding a prolonged bombardment from the frigate. Despite this downbeat review of their situation, all the Marines agreed that they should continue to defend the island, for no matter how small and insignificant South Georgia was to the British Crown, the island was nevertheless a British settlement and therefore entitled to be defended, regardless of the risk. As one of them commented at the time, 'We must not forget that Royal Marine Commandos never surrender unless we are ordered to – and no one has given us that command.'

The battle for South Georgia continued. The *Guerrico* went on pounding the Marines' positions, but as these were well fortified the shells were unable to cause much damage. Occasionally the British would fire back, simply to show they were still alive and well and capable of defending themselves, but their rounds fell harmlessly into the sea, well short of their target.

The following morning the Marine on guard duty reported to Lieutenant Mills that he had seen another ship approaching South Georgia but had not yet been able to identify whether it was friend or foe. Mills flashed a message to Major Norman, the Royal Marine Commanding Officer on the Falklands, saying that he would keep him informed of developments. Norman learnt during that wireless communication that the Argentines had landed that morning not far from Port Stanley and the small force of Marines guarding the Falkland Islands was under serious attack from a large force of Argentine infantry.

As the unidentified ship approached the *Guerrico*, however, the

Argentine frigate made no move and the Marines were certain that the ship steaming towards them must be another Argentine naval vessel. They were right. Within a couple of hours the *Guerrico* had been joined by an Argentine landing ship, the *Bahia Paraiso*. It was obvious that the Argentines meant business.

Through their binoculars the Marines could see that the newly arrived ship was crammed with Argentine infantry. They were aware that the Argentines had first-class Special Forces of their own. One of their crack units was the Buzo Táctico, a Commando-trained assault company of the Argentine Navy with a formidable reputation as an elite fighting force. The Marines were fully aware that if the Buzo Táctico were to take part in a landing on South Georgia they would have their work cut out to defend the place.

Indeed all the Argentine services – army, navy, air force and the coastguards – include Special Force units, which come under the umbrella of an organisation called Halcyon Eight. These units were trained in the 1970s by US Special Forces instructors and Israeli commandos to prepare for the counter-terrorism which Argentina feared so much because left-wing terrorist activity was taking place in many other Central American countries at that time.

Shortly after the *Bahia Paraiso* had drawn near to the *Guerrico* a military helicopter took off and flew straight towards the Marines' position. It was the opportunity Mills had been waiting for, as it allowed him to demonstrate to the senior Argentine naval officers on board the two ships that the Marines were in no mood to roll over and surrender. His men held their fire, enticing the helicopter to come ever closer to their position so that the crew could get a better view of the defences and maybe get an idea of the number of British troops defending the island. When the helicopter flew almost overhead Mills gave the order to fire and his men opened up with everything – rifles, light machine guns and their 66mm rockets – hitting it and bringing it down in the sea. There was tremendous jubilation among the Marines, but shock and horror among the watching Argentines.

Yet Mills and his men were under no illusion that their tiny force could defend the island for long against such overwhelmingly superior forces and firepower, and he reported the situation back to headquarters at Port Stanley. The decision was taken that there was no point in sacrificing the lives of twenty-two Marines and risking the lives of the British men, women and children on South Georgia, who would also be dragged into the fighting. He was ordered to surrender.

Meanwhile the desperate battle to defend Port Stanley continued. The Argentines had taken no chances. They had sent in some six hundred men of the 25th Infantry Regiment, plus the Buzo Táctico assault force. They had come ashore during the night of April 1–2 near Port Stanley in amphibious tanks, which were well equipped to cope with the medium machine guns which were the principal weapons the Marines had for defending the Falkland Islands.

But the Royal Marines at Port Stanley were waiting as the Buzo Táctico launched its first ground attack, aiming to seize control of their barracks. As the Argentine Special Forces soldiers raced towards the barracks, throwing grenades and firing light machine guns, they were met by devastating fire from the Royal Marines' medium machine guns in their defensive positions. The Buzo Táctico realised its battle plan had been too ambitious and withdrew to safe ground.

The Argentines' next plan was to capture the Governor-General and force him to surrender the island and order the British troops to lay down their arms or, if he refused, take him back to Buenos Aires. But the six Royal Marines who had been dispatched to protect the Governor-General and his family were in position as the Argentine snatch squad ran towards Government House. Highly disciplined, they waited until the snatch squad was in the open, running through the garden of Government House towards the residence, before opening fire, killing three of the attackers in the first burst. The other members of the snatch squad ran for cover and withdrew after a further gun battle in which no one appeared to have been injured or killed.

When the attack began the Marines had some sixty men to guard Port Stanley against the Argentines' six hundred infantry soldiers and the Buzo Táctico. At daylight they could see that landing craft filled with infantrymen were being ferried from Argentine ships standing at the entrance to the harbour towards the beaches around Port Stanley. The Marines had set up 66mm rocket-launchers to defend the harbour and these were constantly rocketing the landing craft in an effort to make them head back towards the Argentine ships.

The Argentine forces were firing light machine guns and sub-machine guns as they made their way from the beaches towards the town. They were also hurling grenades left, right and centre, frightening the inhabitants, who had never experienced such an attack in the sleepy life they had always led.

The Marines were now having to pull back to better defensive positions or risk being killed or captured. But they all realised that against such an overwhelming force they had little chance of victory once the Argentines had succeeded in securing one or two beachheads. Within hours of the original assault Major Norman knew that he would be forced to surrender or quit Port Stanley with his forces and make a stand in the rocky hinterland of one of the Falkland Islands.

Norman discussed the position with the Governor-General, telling him that defending Government House to the last man would risk the lives of everyone in the building as well as putting at risk the lives of other Falkland Islanders. He was keen to withdraw his sixty men inland so that he might find a good defensive position which could then act as a beachhead for a future British force trying to wrest back the islands from Argentine control.

The Marines were now fighting at a massive disadvantage, for the Argentines had something like twenty-to-one superiority over the small British force. Then, three hours after the surprise attack, the Governor-General ordered them to cease fire as he surrendered the island to the Argentine invaders.

Back in London, Prime Minister Margaret Thatcher viewed the Argentine invasion as a direct affront to the British people, the Falkland residents and the authority of the British government. She decided to take action despite the fact that since World War Two the British Foreign Office had quietly been suggesting through diplomatic channels that the Falkland Islands could one day be returned to Argentina. It had been further suggested that if Argentina did in fact move forces into the Falklands they would have to do so with the minimum of force and ensure that no casualties whatsoever would occur among the islanders. And that is what had happened, for no British soldier or islander was killed or wounded during the Argentine invasion. Indeed three months before the invasion the British military attaché in Buenos Aires had reported to the Foreign Office in London that the Argentine President, General Leopoldo Galtieri, intended to invade the Falklands, remove the small British force and reclaim the islands for Argentina. The diplomat even gave the exact date of the invasion. No one in London had taken the slightest notice of the warning.

Margaret Thatcher, however, had not been privy to the reports flowing into the Foreign Office, nor to the Foreign Office's diplomacy. She wanted the Falklands returned to British sovereignty immediately and for Argentina to withdraw all its troops and hand back authority to the Governor-General. The Argentine government refused and Thatcher decided that Britain would retake the islands by force.

Initially, the military advice given to the Prime Minister by senior British military and naval commanders was decidedly downbeat. They pointed out that the Falklands were thousands of miles from Britain in the middle of the South Atlantic, with no friendly countries or bases where British planes could refuel. They also pointed out the logistical nightmare of, firstly, moving sufficient numbers of assault troops to the islands quickly enough and, secondly, re-supplying and reinforcing them if these measures should prove necessary. It didn't help that the Tory government had mothballed five Royal Navy frigates because, as they themselves explained, there wasn't enough cash in the defence budget to pay the oil bills! Nor did the senior military officers believe that the Argentine Army would be a pushover.

But Thatcher would hear none of it and resolved to press on with her decision to recapture the Falklands. To many it seemed that she was taking an incredible political risk, for failure would more than likely end in her being forced to resign as leader of the Conservative Party. Many were also of the belief that she was trying to emulate, in some small way, the achievements of her great hero, Sir Winston Churchill. In 1982 she had not yet stamped her authority on her own party, the House of Commons or the British electorate. That would come much later and it would be given a great push by the fact that she had had the courage to risk all to retake the Falklands Islands against considerable odds.

But Thatcher was lucky that in the British armed forces, and in the Special Forces in particular, she had men of great calibre, ability and courage. She put her future in the hands of those men and they did not let her down, though there were times when she must have wondered whether she had made the right decision.

The senior British commanders told her bluntly that there was no chance of a successful outcome to the suggested military operation without help from the United States. She talked to her new-found friend, President Ronald Reagan, who had only been in office some fifteen months, and initially he showed little enthusiasm for the idea. His State Department reported to him that Britain's Foreign Office had for years been trying to find an amicable way of getting rid of the Falklands and permitting the Argentines to take over the place.

At first Reagan tried to persuade Thatcher to drop the idea, but she would have none of it. Britain needed assistance in two vital areas to give the plan any chance of success. It had little knowledge of the current political nuances in South America, for since World War Two the Americans had taken the continent as its sphere of influence. And it had virtually no MI6 contacts in Argentina or its neighbours and would need to rely on America's CIA for such assistance and information.

More importantly, Britain would require military help from the United States. It would need air-to-air refuelling aircraft so that British warplanes could fly to Ascension Island, a British base in the Atlantic, refuel and then head for the Falklands. The British bombers had enough fuel for the flight south but not for the return. The only nation with sufficient refuelling aircraft was the United States.

Britain also needed America's help to ascertain the Argentine positions and defences on the Falklands – information which could only be provided by US spy planes. Britain had none. Without such aerial photographs, the naval and military planners would be at a severe disadvantage.

Eventually, after much persuasion, Reagan agreed to help but insisted that if the British naval or military forces ran into trouble or faced defeat the United States would not come to their rescue. The buck would have to stop at 10 Downing Street.

The British public were given little or no information about the extraordinary behind-the-scenes efforts that were being made by the civil service, the Ministry of Defence and the armed services to put the entire operation on the road as speedily as possible. The

Special Forces demonstrated their state of readiness and speed of action faster than any other branch of the services. Within three days of Argentina's invasion of the Falklands, 3 Commando Brigade, Royal Marines, had sailed from Portsmouth with the first elements of the Task Force.

But it would fall to the Special Boat Service to take many of the honours in the first phase of the action. Twenty-four hours after the first reports of Argentine forces invading South Georgia, 42 Commando, SBS, who had just returned to Poole, in Dorset, from three months' winter training in Norway, were put on standby and immediately moved to RAF Lyneham, in Wiltshire, where they boarded a flight to Ascension Island. Shortly after arriving there they were joined by SAS D Squadron, making a total force of some fifty men. Later they would be joined by M Company of 42 Royal Marines. Their objective – to reclaim South Georgia.

In stormy weather and blizzard conditions three Wessex 5 helicopters, with forty SBS and SAS special forces on board, headed for Possession Bay. But as they reached the island they hit a wall of snow. Visibility dropped to zero and the three choppers had to return to the destroyer HMS *Antrim*. Three hours later they set off again and, after landing precariously on a glacier, the men managed to struggle out of the choppers with their three pre-loaded sleds.

They managed to stay alive during that night but conditions had become intolerable, as they were suffering hypothermia and their sub-machine guns didn't work in the intense cold. The *Antrim* sent helicopters to the rescue. However, shortly after picking up the frozen men the lead chopper hit a white-out and crash-landed at about forty miles per hour on the glacier. Everyone on board was thrown around, but no one was killed or even injured. The smashed Wessex was a write-off. Having ditched most of their equipment, the troops clambered on board the other chopper and set off once more. Again a white-out descended and the helicopter crash-landed on the ice at some seventy miles per hour, ending up as a tangled mass of metal. Incredibly, the troops crammed together inside, some thirty-two men, survived, though some suffered injuries.

That night the men on South Georgia had to stay huddled close together in freezing conditions as the snow and ice closed in around them. They had hard rations, but by the morning some of the men were all but frozen. The other chopper waited for a break in the weather and then flew to South Georgia, landing near the smoke marker the men had sent up. The men clambered on board, leaving all their equipment behind. They made it back to the *Antrim* and somehow the chopper and the men survived yet another crash-landing on the deck. After three days of frustrating work in horrendous conditions, nothing whatsoever had been achieved.

It was decided to abandon attempts to chopper troops on to the storm-lashed island from a distance. The ice-patrol vessel HMS *Endurance*, which had been operating in the South Atlantic for some months, sailed into Hound Bay before the SBS were airlifted in small Wasp helicopters on to the ice. Other SBS men took to the water in their small inflatable craft, fitted with powerful outboard engines, with the necessary stores, equipment and ammunition. Some hours later they made contact with each other and finally set off on their mission to find the Argentine troops, recce the place and report back to headquarters on board the *Antrim*.

Their target was King Edward Point, where the Argentines were believed to be putting ashore more troops and supplies, but to reach that area would mean travelling some fifteen miles across glaciers and taking an eight-mile boat ride across Cumberland East Bay. And all this in horrendous winter conditions of blinding snow storms, white-outs, freezing temperatures and howling winds. When they arrived at the edge of the Bay they found the sea frozen over. The SBS team radioed the *Antrim* to no avail. They tried other radio frequencies in a bid to contact other Royal Navy ships in the area, but there was no reply. It was only later that the SBS learnt that all British ships around South Georgia had been ordered out of the area after British Intelligence discovered that the Argentine attack submarine *Santa Fe* was patrolling those seas. Unbelievably, no one had informed the SBS what was happening. Two days later it was decided to abort the SBS mission and the troops were airlifted off by the Wasp helicopters and returned to the *Antrim*.

On April 25 it was decided that another SBS unit should make a further attempt to land on South Georgia with the same intention of getting as close as possible to the Argentine headquarters and relaying to the *Antrim* as much information as possible.

Having put the last SBS men on to South Georgia, the Wasp pilot Lieutenant Ian Stanley spotted the *Santa Fe* on the surface near Cumberland East Bay. He swooped on the submarine and dropped his six depth charges around the vessel. It seems they must have done sufficient damage to prevent the submarine diving beneath the waters. Wasps from HMS *Endurance* and a Lynx chopper from the frigate HMS *Brilliant* were scrambled and blasted the *Santa Fe* with missiles and machine-gun fire. The stricken submarine limped away.

Of course, this assault meant that the planned element of surprise had been sacrificed, so the decision was taken to mount an attack as speedily as possible, before the Argentines had time to organise proper defences. All British Special Forces on the ships then around South Georgia were called together and ordered to carry out an attack a.s.a.p on the Argentine headquarters at King Edward Point. There were just seventy-five men, about one-third of the number of Argentines defending the HQ.

During a break in the hellish weather, the British Special Forces contingent, bristling for a fight, were choppered on to South Georgia and made for Grytviken. When the small, elite band finally arrived on a mountain ridge overlooking the town they were stunned to see Argentine troops laying out white bedsheets, which would have been clearly visible from the air.

In battle formation, the British soldiers set off for the town with their weapons at the ready, just in case they were walking into a trap. When they arrived they saw some two hundred Argentine soldiers formed up beside their national flag, as though on parade, and an officer stepped forward and formally surrendered. But the Brits were taking nothing for granted. The Argentine forces were ordered to lay down their weapons and these were collected by the British troops and taken to one side while the Argentine forces remained at attention. The British officers knew that they were still outnumbered and needed reinforcements to be airlifted in as quickly as possible. Those on board the *Antrim* and the other Royal Navy ships in the area were staggered that the small band of Special Forces men had accomplished their mission so quickly. More troops were immediately dispatched to Grytviken and news of the victory was flashed to London and passed on to 10 Downing Street. 'South George Recaptured', screamed the newspaper headlines.

After the Falklands had been retaken it was learnt in Buenos Aires that the reason why the Argentines had surrendered so meekly and so quickly, without a fight, was that senior Argentine officers had heard that Britain had dispatched its Special Forces to recapture South Georgia. They knew the game was up before one shot had been fired.

Now the SAS and the SBS were immediately switched to the Falkland Islands, for their expertise and efficiency would be needed as Britain speedily mounted its largest military operation since the Suez landings of 1956. Sixty-five warships, crammed with soldiers and Special Forces troops, supplies, weapons, tons of ammunition and everything required to carry out a successful major assault against a well-equipped and large enemy force, steamed to the South Atlantic at full speed.

United States Intelligence had reported that the Argentines were pouring more men, supplies and heavy weapons into Port Stanley.

Though they had surrendered South Georgia without a fight, it seemed that they were determined to hang on to the far more important Falkland Islands.

As the large British force made its way towards the Falklands, men of the SBS were already on the islands reconnoitring potential landing sites for an invasion, watching Argentine forces preparing defences and relaying information about Argentine warships and transport planes which were sending in supplies to their troops.

There was one major problem in sending back information to the senior officers of the Task Force. Any radio link would be intercepted by the Argentine signals squad, revealing that the British already had troops on the islands. Knowing this, the SBS supplied all the intelligence by means of the much slower Morse Code. Sending complicated recce reports was impossible with Morse. In such cases SBS men had to be picked up by Sea King Mark 4 helicopters or taken off beaches by small boats, and then taken to a British warship offshore, where they explained the reports before being returned to the islands. Nonetheless, the system worked satisfactorily enough.

SBS recce teams of four men were dispatched to eight main areas surrounding Port Stanley, including San Carlos, Eagle Hill and Ajax Bay. Sometimes the SBS patrols had to hide out for two weeks at a stretch, surviving in 'fox holes' covered by turf while watching the Argentine forces.

They were receiving reports of the war from those mates returning from briefing sessions. In this was they heard of the sinking of the Argentine *Belgrano*; the Argentine Exocet missiles playing havoc with the Royal Navy ships; and the sinking of HMS *Sheffield* with the loss of twenty-one men.

One brilliant SBS operation involved eighteen soldiers of 2 SBS and 6 SBS, ordered to capture the *Narwal,* an Argentine fish-factory ship which was in the British exclusion zone around the Falklands. As the men were approaching the thirteen hundred-

A crack shot from the French Foreign Legion.

Top: The newly-formed Special Air Service undergoing pre-parachute training by leaping off the tailgate of moving lorries at No 2 Parachute Training School, RAF Kabrit in November 1941.

Bottom: Frogmen from the Royal Marines Special Boat Service boarding an RAF Hastings in 1952 for a parachute insertion into the English Channel

US Rangers from 75th Ranger Regiment, armed with Colt M4 5.56mm rifles fitted with M203 40mm grenade launcher and laser sights, during a night attack at Camp Merrill, Georgia.

Inset: A US Ranger's badge from the 10th US Special Forces Group.

The SAS storm the Iranian Embassy in 1980. Nineteen hostages were rescued, and three gunmen shot dead.

Top: A patrol from 22nd Special Air Service Regiment form up near a jungle Drop Zone off the Bentong Road, Malaya, in December 1952 during operations against the communist terrorists during the Emergency.

Bottom: Two soldiers from the modern-day 22 SAS, armed with M16A2 5.56mm rifles with M203 40mm grenade launchers fitted, undergoing continuation training in the jungle in Brunei.

Top: Septsnaz Spetsrota counter-terrorist soldiers from the Felixdzerzhinsky Division, armed with Kalashnikov automatic weapons, cover team members as they force 'prisoners' up against a wall and search them for weapons during an exercise in the unit's training village.

Bottom: US Marines break a door in a deserted school as they search for snipers in Mogadishu, Somalia.

A soldier from Pathfinder Platoon, 16 Air Assault Brigade, armed with an M16A2 5.56mm rifle fitted with an M203 40mm grenade launcher, is silhouetted against the last glow of daylight in the Jordan desert whilst training with the Jordanian special forces. This is the weapon used by Chris Ryan, the sole member of the SAS patrol Bravo Two Zero to escape from Iraq to Syria during the Gulf War.

The devastation of the bombing of the Grand Hotel in Brighton; one of the members of the Black September terrorist group on the balcony of Israeli House at the Munich Olympics; and the terrible vision of the burning of the Twin Towers in New York on September 11th. It is to counter events such as this that the world's highly-trained special forces exist...

tonner in Sea King helicopters two Harrier jump jets roared overhead and bombed and strafed the ship. The Sea Kings hovered over the damaged ship, the men fast-roping down to the deck, guns at the ready. But the men on board the *Narwal* put up no fight and the SBS rescued all the crew, took possession of charts and operational orders from the Argentine naval headquarters, and then placed charges which exploded minutes after they had been airlifted from the deck. It was a classic mission.

But the real battle was yet to take place.

Days before the British troop landings, 2 SBS and 3 SBS, with an SAS mortar detachment, were choppered from the *Antrim* to San Carlos Bay armed with a new device never before used in combat, a thermal imager, which can pick up the presence of people simply from body-generated heat.

Around San Carlos the thermal imager picked up a signal revealing a company of some thirty Argentine soldiers in camouflage hiding out of view of the naked eye. The information was flashed back to the *Antrim*, which bombarded the area with its 4.5 inch guns. Then the Special Forces team went in as the Argentines opened up in an attempt to shoot the Royal Navy choppers out of the sky. Somehow they landed virtually unscathed. Using a loudhailer, a British officer called in Spanish for the Argentines to surrender, but they replied with machine-gun fire. The officer tried again, and this too was met with a burst of fire.

'We're going in' were the only words the Special Forces men needed to hear and a plan of attack was quickly put into action. It was in fact a classic flanking movement, with some SBS keeping the Argentines busy firing from their positions while other SBS and SAS troops retreated under cover and then carried out a flanking movement before taking the enemy by surprise with withering bursts of fire from close range. Twelve Argentines were shot dead, three were wounded and nine were taken prisoner. That attack might well have saved the lives of many on board the British warships because the Argentines had been manning anti-tank guns and mortars covering the straits of San Carlos, into which the British warships were about to enter.

Under cover the SBS moved down to nearby Fanning Head, where they knew the Parachute Regiment would storm ashore at zero hour unseen by the sixty Argentine soldiers who were sheltering in houses in the Port San Carlos area. The landing went without a hitch and as soon as the Argentines saw the strength of the British invasion force they moved out of the houses. A few of them, however, took up defensive positions and shot down two Royal Marine Gazelle helicopters, killing three men on board.

The SBS were involved in nearly every landing by the British forces, making recces, reporting back, watching the Argentine forces and waiting at the invasion site when the British troops stormed ashore. It was classic seaborne Special Forces action, always at the sharp end, always taking risks, always keeping the invasion force informed of what to expect. It meant that no British force landed on either East or West Falkland without the highest-quality intelligence, fully aware of enemy positions and strengths. That intelligence saved many British lives.

As the Parachute Regiment moved inland to confront the Argentine forces on East Falkland, the SBS were moved to the larger West Falkland, where most of the islands' population lived, where the Governor lived and from where the Falklands as a whole were administered. SBS units were secretly deployed in ten locations, to send back intelligence reports about the enemy and provide recce information for potential landing sites. The SBS also directed some of the vital naval bombardments after moving into exposed positions only one thousand metres from the enemy targets.

Before the main invasion the SAS conducted a brilliant operation on Pebble Island, off the coast of West Falkland, where the Argentine engineers were preparing an airstrip for Pucara ground-attack fighters and constructing a rudimentary radar station. Four SAS men of D Squadron of the Mountain Troop were inserted on to the island by the Boat Troop and reported back that eleven Argentine planes and a quantity of bombs, ammunition and stores were on the airfield and open to attack. The go-ahead was given and the SAS were tasked to carry out the operation.

The four-man Boat Troop choppered into West Falkland again and set up an observation post overlooking Pebble Island. After dark they launched their silent canoes, landed on the island before midnight and set up a radio link with HMS *Hermes*. The second four-man SAS unit moved up to towards the airstrip and, finding a safe hideout, laid low until daybreak. Having recced the airstrip in great detail, the SAS men stayed under cover until nightfall and then returned to the position of the Boat Troop, who radioed the information back to the *Hermes*.

It was decided the entire D Squadron should take part and they were flown in three helicopters from the *Hermes*, flying just above sea level without lights in darkness. As the troops landed on Pebble Island the moon came out and after a fast march the Squadron arrived in position at 7 am.

The attack began with small-arms fire and rockets which destroyed several of the Argentine aircraft before those manning the garrison realised they were under attack. Within sixty seconds of the SAS opening fire, HMS *Glamorgan* poured shells non-stop on to the airstrip and the infantry defences. This heavy firepower enabled the SAS units to move on to the airstrip and set explosives under every aircraft, the ammunition and bomb dumps, the petrol tanks and the stores. Within fifteen minutes of the start of the attack, the mission was over, the airfield, the supplies and all the planes destroyed.

The SAS, with only one man wounded, beat a retreat and were picked up two hours later by helicopters and returned in triumph to *Hermes*. As news spread through the ship of the destruction caused by the SAS, those on board queued to congratulate them. 'All in a day's work,' sums up their cool reply.

The last SAS attack of the Falklands War, a seaborne raid on the oil tanks near Stanley Harbour, was carried out in conjunction with the SBS. Four fast Rigid Raider assault craft, manned by the SBS and filled with SAS and SBS men, were ordered to cause mayhem in setting the oil tanks ablaze to divert attention from 2 Parachute Regiment, who were about to launch an attack on the all-important Wireless Ridge, from which the Argentine guns had an overall command of Port Stanley.

But this time there was no quick victory for Britain's crack Special Forces. As the Rigid Raiders entered the bay in darkness and sped across the water to the oil tanks, an Argentine hospital ship anchored there turned its searchlights on them. Brightly illuminated for the Argentine forces, the four boats came under heavy and prolonged fire from medium machine guns, which was ripping holes in their sides and threatening to sink them.

An urgent wireless message was flashed to Major General Julian Thompson, of 3 Commando Brigade, the cutting edge of the Task Force, seeking assistance. The men in the Rigid Raiders were taking a hell of a battering and unless help came soon there was every probability that they would need to withdraw. They urged that 2 Para be diverted from their planned attack on Wireless Ridge to take out the medium machine guns pinning them down in the harbour.

'Forget it,' came the reply from Thompson. In essence, he told them, 'If you're big enough to get into the shit, you're big enough to get out of it.'

The SAS and their SBS colleagues had no option but to turn and run for it – back to base. But they didn't like the idea of having to back off so ignominiously. Indeed some of the men were furious that Thompson had refused their request to send 2 Para to the rescue so that they could finish their task of destroying the oil tanks. However, some senior planners back in London were rather pleased with the failure of the mission, for the British forces – the Royal Navy, the Army and the RAF – as well as the Falkland Islanders, needed the fuel.

The Falklands War, wrapped up very satisfactorily within ten weeks of the Argentines' seizure of the islands, was a great tribute to the planners and the professionalism of Britain's armed forces. The British forces lost two hundred and twenty-five men and more than seven hundred were wounded. As Robin Neillands wrote in his book *In The Combat Zone*:

Few complaints have come from the men who fought in the Falklands Islands. The war was won, at no great cost in lives, and when all the risks are taken into account that alone is a minor miracle. The main contribution of Special Forces and Special Operation Forces to the Falklands War was in the provision of first-class, well-trained troops, who could fight and win against heavy odds and in inhospitable terrain at the end of a long logistical trail. Without such troops, and a knowledge of their abilities, the retaking of the Falkland Islands would have been impossible and could not have been seriously contemplated.

There are other Special Forces around the world which train soldiers for similar duties to those of Britain's SBS. And among the best trained and best equipped are the famous United States Navy SEAL (Sea, Air, Land) force, who can rightly boast of being a legendary body of professionals.

The United States Naval Special Warfare Command trains men for five dedicated types of operations. These include short, direct-action missions such as launching attacks against ships at sea or destroying facilities afloat or on the shoreline. This section also includes small, short, sharp combat operations against enemy forces. Special reconnaissance and surveillance operations, covert beach surveys and observation-post missions are other elements of the special training.

The SEALs also train for more dangerous operations – for example, equipping, training and leading friendly guerrilla forces behind enemy lines to make contact and conduct small-scale firefights with enemy forces. Since the attack on New York's World Trade Center in September 2001, counter-terrorist training has become increasingly important. Great emphasis is now placed on training for one-off missions, as well as for long, hard counter-terrorist operations against a determined, well-armed, well-disciplined enemy which may continue for months.

And SEALs are also trained to train and advise military, paramilitary and law-enforcement personnel of allied and friendly nations in non-combat roles.

One of the SEALs' classic exploits – code-named Task Unit Whiskey – took place in December 1989, when they were involved with the capture and removal of General Manuel Noriega, leader of the Republic of Panama, from his seat of power. The overall operation, code-named Operation Just Cause, was ordered by President George Bush on the advice of the CIA, because they believed Noriega had become a threat to the stability of Central America and a danger to the United States.

The mission of the team of US Navy SEALs assigned to Task Unit Whiskey was to find and immobilise General Noriega's personal patrol boat, the *Presidente Porras*, which the CIA had identified as his principal plan of escape should the United States forces invade Panama and start looking for him.

Launched from Rodman Naval Station in Balboa Harbour – one of many US naval depots in the Panama Canal area – the twenty-one divers, swimmers and boat men in their two Combat Rubber Raiding Craft (CRRC) headed slowly and silently out to sea just before midnight. The men had back-up in case the Panama Defence Force deployed heavy firepower.

Within range of the port where the *Presidente Porras* was berthed a fire-support team manning .50-calibre heavy machine guns, Mk 19 automatic grenade-launchers and 60mm mortars, all equipped with night sights. Before dark their targets had been selected and the guns trained on the objectives. Two more patrol boats carried more SEALs in case their comrades ran into serious trouble and needed rescuing.

The plan was for the underwater swimmers to be dropped off their patrol boats about half a mile from the target, swim to Noriega's boat, silently place and arm the demolition explosives on the hull and then swim back to their waiting craft without being noticed.

Everything went according to plan except for the fact that the high-powered, high-revving engines weren't designed for slow, gentle travel. The engines began coughing and spluttering, drawing the attention of fishing smacks that were heading out to sea. But they kept going and made it safely to the tree-lined shore without being noticed by the patrol guards on duty.

Then they saw patrol boats crossing the harbour and from that distance they had no idea whether one was Noriega's escape boat making a leisurely dash for the open sea or Panama Defence Force patrol boats checking that no US Navy vessels – big or small – were anywhere in the vicinity.

Then, through their wireless headphones, the SEALs team were advised that the operation would now start thirty minutes earlier than scheduled, which gave them no chance of carrying their carefully prepared plans. But they were ordered to detonate the *Presidente Porras* whatever happened. The only way to blast the explosives on time was to take their CRRCs closer to the target and risk being spotted and shot at. There was no other choice.

The boat men fired the engines but Boat Number 1 refused to start, so they had only one manageable CRRC with which to make good their escape. Commander Norman Carley, the mission commander, took the first boat slowly and quietly nearer the objective and the first dive pair slipped over the sides with their precious cargo of explosives. He then returned to the coast, fixed a tow line and pulled Boat Number 2 out towards the target. The second dive pair disappeared into the water and Carley returned to the protection of shadows on the coastline.

However, on their slow return voyage wireless contact was resumed with the urgent message that the operation had been brought forward another fifteen minutes, which meant that when the attack proper took place by land, sea and air the SEALs swimmers would be just approaching their target. There was nothing that Carley could do and there was no way that he could alert his swimmers that their task was now far more urgent, dangerous and risky.

The dive pairs swam towards the target some twenty feet underwater so that no ripple would appear on the surface. They could not swim too fast, for between them they were carrying a Mk 138 Mod 1 charge loaded with twenty pounds of water-resistant explosive in a haversack, complete with an MCS-1 clock, a Mk 39 safety and arming device and a Mk 96 detonator, all designed to provide reliable delay and detonation to the main charge.

To provide greater security the swimmers used the Draeger re-breathing system, which recycles expelled air, cleans out carbon dioxide and replenishes the oxygen. It is a closed system ensuring no bubbles reach the surface. And, for once, the swimmers discarded the usual high-tech navigation system and instead relied on the old-fashioned method, wearing a luminous compass on the wrist and calculating the distance travelled by counting the number of kicks they made. It worked perfectly.

As the first two swimmers surfaced under Pier 18 the place seemed to explode. To their surprise a firefight was in full progress, something which was not meant to have started until fifteen minutes later. They wondered what the hell had gone wrong, and feared that someone might have tipped off the Panamanian guards that a seaborne attack was about to take place. It indeed appeared that the guards were nervous, for they were firing out to sea and throwing hand grenades into the harbour as if they had been tipped off that US SEAL swimmers were in the vicinity and planning an attack.

By now the first pair of divers were beneath the pier and they could see the *Presidente Porras* still berthed at the dock. Together they dived beneath the boat out of sight of the patrolling guards. They swam to the stern and were in the process of clamping the first explosive on to the port propeller shaft when the engines suddenly and unexpectedly roared into life. Now they really had to work fast.

As the engines chortled away, occasionally being revved up, the two divers attached and armed the charge in less than two minutes. They checked it was ticking and then left as silently and unobtrusively as they had arrived. The patrol boat's engines were still running.

One minute later, and bang on schedule, the second team of divers arrived, and clamped their Mk 138 demolition system to the starboard propeller shaft. They tied a length of detonation cord around the charge already on the port shaft, using a dual-priming technique to ensure that both charges exploded at the same time. The two divers removed the safety catches and started the clock, set to cause the charges to explode forty minutes later.

But before the swimmers could move away from the target boat, the Panamanian soldiers once again began throwing grenades into the water around both the vessel and the pier where it was berthed. Fearing the main charges were about to detonate, the swimmers took refuge behind the pier's pilings. Still the grenades rained down all around them, although they were convinced that the Panamanian soldiers had had no real sighting of them and were only guessing what was about to happen.

When the swimmers realised there were only ten seconds to go they both dived and swam as deep and as fast as possible. At precisely 10 am they heard the two Mk 138 charges explode dead on time, causing chaos and mayhem among the soldiers on the pier. Heads would roll when General Noriega was informed that his pride and joy, his favourite toy, his wonderful fast patrol boat, had been blown up.

The four swimmers teamed up, all reporting no injuries and no problems, and headed back across the bay. Waiting anxiously for them were CRRC boats and very relieved officers who had not been in communication with any of the four men for more than ninety minutes. The swimmers' MX 300 waterproofed radios had not been up to the task. They had heard nothing. Watching the action from a distance, the officers were seriously concerned that the swimmers had not survived the mission. The fact that the main charges had exploded was proof that the teams had done their job despite the firefights and the problems, but from the view they had of the gun battle it was obvious that the Panamanians had put down a lot of metal and grenades into the water.

The task had been a great success, a tribute to the professionalism of the SEALs who had taken part. They had continued the tradition of the United States Navy Underwater Demolition Units, which had been established during World War Two. These cleared obstacles from enemy beaches, took out beach OPs, targeted beach defences and supplied valuable intelligence about enemy positions, numbers and guns on the shorelines of Europe and the Pacific. Theirs were all highly dangerous missions which have been replicated on many occasions during the past sixty years.

But what is special, almost unique, about the SEALs and their fellow seaborne Special Forces is that the men who volunteer and take part in such operations are quiet, confident, determined and cool-headed; men who don't boast of their exploits or even talk much about the extraordinary, often life-threatening, risks they take to ensure that a mission is accomplished.

CHAPTER 7

22 SAS

O NE BEAUTIFUL MAY EVENING a yellow JCB digger
bounced gently along the narrow, hedge-lined lanes of
County Armagh. The birds were singing and chirping in the trees,
which hadn't yet come into full leaf, and the road between
Armagh town and Portadown was almost empty.

As the JCB approached Loughgall, a quiet, picturesque village
with some two hundred and fifty people, a couple of pubs, a
police station, a school, a post office and a few shops, a van drove
up behind the digger and was waved past. The small blue Toyota
Hiace overtook the JCB and headed towards the deserted police
station a couple of hundred yards down the road.

This peaceful scene took place in 1987, during the height of the
troubles in Northern Ireland. Using the bomb and the bullet, the
Provisional IRA were waging a war of terror against the ruling
Protestant majority, the hated Royal Ulster Constabulary, the
British Army and the SAS. They had already killed hundreds of
innocent people, wrecking town centres with their bombs and
terrorising country areas in their determination to drive out
Protestant farmers and smallholders.

It was because of IRA bomb attacks on the sergeant and his two
constables that Loughgall's police station was closed and had

been empty for months. Indeed this Protestant village was a far from thriving community, most of the young people having gone to find work in Belfast or mainland Britain.

As the Hiace passed the JCB, two men wearing dark blue boiler suits, one seated each side of the driver of the digger, pulled black balaclavas over their heads, stood up at either side of the cabin and picked up Kalashnikov AK47 automatic rifles. In the bucket of the JCB was a massive bomb.

When the van reached the police station it came to a halt. On catching up, the JCB slowed down to a walking pace and turned towards the building. But it didn't stop. Instead it accelerated and smashed through the wire perimeter fence, with the two men still holding on to the cabin, and continued to trundle on towards the police station.

Seconds before the JCB crashed into the white, two-storey building, all three men jumped out and began to run back through the gap in the perimeter fence made by the digger. At the same time the back doors of the van opened and six more men dressed in balaclavas and dark blue boiler suits, and carrying handguns or rifles, jumped out on to the road.

They were met with a hail of automatic fire from the hedgerow on the far side of the road. All along the hedge bursts of fire could be seen. The three men from the JCB turned and fled in terror along the road they had driven down. Those who had leapt from the van reacted in different ways. One or two froze and were shot dead, others chased their comrades down the road, while others threw themselves behind and underneath the van in a bid to escape the devastating hail of bullets.

Those behind the van returned fire as best they could, but they had no clear targets to aim at.

All of a sudden a massive explosion filled the air as the bomb ripped the roof off the police station and splintered wood, tiles and masonry showered down over the whole area. The

Provisional IRA men knew their only chance of survival was to stay behind the cover of the van and take potshots at those who were still keeping up heavy fire. One or two of the Provos were hit in the shins and the feet by bullets ricocheting off the ground and the van took a hammering from the ferocious onslaught.

After a few minutes the IRA gunmen found themselves under attack from another angle and this time there was nowhere to take cover. They were sitting ducks. Four went down in two further bursts of fire, but still the others were shooting at their attackers But not for long. The three Provos still capable of running decided to make a dash for freedom and took off from behind the van. They had run about a hundred yards along the road before they were met by a hail of bullets which cut them down.

'Cease fire,' came the order in a clear, crisp and loud voice as silence descended on the grotesque scene. In the background the police station was burning furiously, flames and smoke billowing some twenty feet into the evening air. Around the Hiace van a number of broken bodies lay in contorted positions, and along the road were more, all with rifles close to them.

From the hedgerows emerged a dozen soldiers in full combat gear, their faces darkened, their green-and-brown camouflage fatigues blending with the background, their helmets covered in hessian, and all carrying sub-machine guns, rifles or machine pistols. But none of them relaxed. Two checked the bodies for signs of life while others stood by, their guns pointing at the prone bodies just in case one was still alive. But all were dead.

The firefight had lasted exactly ten minutes, during which eight Provisional IRA gunmen were killed. The Special Forces operation also provided police forensic teams with the weapons the Provos had been using that night. These had been used in seven murders and nine attempted killings. One .357 Luger magnum revolver had been taken from the body of a part-time officer of the Royal Ulster Constabulary who had been shot dead during a gun and bomb attack on the RUC station at Ballygawley, in County Tyrone, in December 1985. The same handgun had also been used in three further killings and an attempted murder. The other weapons recovered from the dead men had all been used by the IRA in other killings, and included three Heckler & Koch rifles, a Belgian FN rifle, two .223 FN rifles and a twelve-bore shotgun.

Among the Provos killed that night were two prominent members: Patrick Kelly, the Commanding Officer of the East Tyrone Brigade, a thirty-year-old married man with three children, and Jim Lynagh, thirty-two, a former Sinn Fein councillor for Monaghan. The East Tyrone Brigade had been one of the most active IRA cells in the Province for the past ten years. They launched the campaign to destroy local police stations in villages and outlying districts in their area and would intimidate builders brought in to repair damaged public buildings, threatening to kill them and their families if they carried out any such work.

There had been another particularly evil and sinister side to the East Tyrone Brigade's *modus operandi*. For some seven years they had been deliberately targeting and killing lone Protestant families with only one son. Several farmers whose forebears had worked their land for generations had been shot at their doors, unarmed and often in full view of their families, leading some Unionists to claim, with some justification, that the IRA was waging a genocidal campaign against Protestant families.

The SAS is considered to be one of the best, if not the very best, Special Force in the world. Most Special Forces are based on the training and organisation of what is properly called the 22nd Regiment, Special Air Service, or 22 SAS, but it is the professionalism, discipline, camaraderie, courage and determination of this British force that maintains it in such high regard among its peers throughout the world.

The operation described above is meat and drink to the men of 22 SAS but nonetheless it was carried out with ruthless efficiency and not one SAS soldier was killed or wounded. In the Mess that night the team enjoyed a few beers together and celebrated a job well done. They didn't gloat about killing eight Provo gunmen and bombers but were simply pleased that everything had gone according to plan and that the objective had been achieved with no casualties to the unit.

During that ambush in Provo country the twelve SAS men adopted the same procedures and disciplines as if they had been on a mission behind enemy lines in a foreign country against large and formidable enemy forces. The SAS is trained to take the war into the enemy camp. First, a small advance party is inserted behind enemy lines to identify the problems and search for a secure, concealed forward base which is close to a water source and a suitable DZ, or dropping zone. Only when that has been achieved will the rest of the SAS team move forward into the delegated operational area.

The smaller units inserted deeper into enemy-controlled areas are expected to be totally self-sufficient and, if the operation is to extend over several weeks, they will have to survive by living off the land. On most occasions the unit will also be responsible for ensuring they can extricate themselves to a prearranged pick-up zone and not necessarily rely on being taken out by helicopter. These operations may involve only surveillance or reconnaissance; others may involve military action.

The SAS deploys only four fighting units, called Sabre Squadrons, each of which is divided into four troops, each of sixteen men. Each troop consists of four independent patrols, each manned by four cross-trained specialists. Typically, these four-man patrols, the SAS 'bricks', are used during offensive missions against a major enemy behind their lines. They will be tasked with extended surveillance of vital targets such as military formations or movements, roads, railways, airfields, shipping lanes, and this information must be passed back as soon as possible. Today there is also a wide range of electronic devices which the SAS use to watch and listen to the enemy.

When the bricks arrive at an ideal observation post, or OP, which the SAS calls a 'hide', one man will be tasked with keeping watch while the other three construct the hide. Working behind a hessian screen, these three clear the immediate area and set up a camouflaged, waterproof covering as protection against enemy aircraft as well as the weather. Then they camouflage the hide with fallen branches and other suitable greenery, taking care not to disturb the surrounding undergrowth or foliage or cutting any branches or twigs. Most hides are shaped like a cross, each soldier occupying one arm of the cross, with their kit in the centre. This cross-shaped form provides all-round observation, is easy and quick to build and offers a number of escape routes. At all times one man is on sentry duty while another is responsible for radio communications and helps to keep watch, while the third man attends to personal needs and rests and the fourth man sleeps. The men rotate their positions anti-clockwise at hourly intervals to ensure that the sentry is replaced before he becomes tired.

Several factors have to be considered before the final decision is taken on exactly where a hide will be built. There must be a nearby source of water and the position must offer unrestricted observation and radio communications. Also, the hide should be sited away from roads, tracks, buildings or anywhere else that might be discovered accidentally.

Today a brick's intelligence-gathering activities usually extend far beyond the normal range of vision, thanks to the latest electronic gadgetry, which can be used to monitor roads, railways, dead ground or areas of low visibility. Unattended ground sensors (UGS) are also much in use nowadays. Magnetic sensors can detect ferrous metals used in small arms, vehicles, tanks, artillery and other military hardware, all of which distort the earth's magnetic field.

Other remote electronic devices used in surveillance work include Audio Unit DT-383/GSQ, an integral microphone which can detect noises with the sensitivity and frequency range of the human ear; AN/GSQ-160, a disposable seismic intrusion detector which emits a continuous radio frequency signal that senses changes in the reflected energy caused by vehicles or personnel up to one hundred and twenty feet away; AN/GSQ-160, an electromagnetic intrusion detector which radiates a continuous radio signal on two frequencies and senses changes in reflected energy resulting from movement; AN/GSQ-176, an air-delivered non-recoverable seismic intrusion detector which looks like a small tree and transmits a radio alarm signal to circling warplanes, pinpointing the bombing target; and AN/GSQ-154, a miniature seismic detector equipped with a logic unit and external geophone which minimises background signals and issues a radio alarm signal when intrusion is detected by vehicles or ground troops.

There are also other magnetic sensors and intrusion detectors which inform those in the hide that there are either vehicles as far away as seventy-five feet or humans as close as twelve feet. Seismic sensors, which detect vibrations, can be used either to protect the hide or to gather intelligence. There are acoustic sensors which, in effect, are highly sensitive microphones; disturbance sensors, which transmit a sweeping radio alarm when moved or stepped on; and, infrared sensors, usually employed to defend large perimeter areas.

Different surveillance techniques are employed by SAS soldiers who are using their hides for longer periods of time. Sometimes a brick has to watch a vast area of countryside and collect accurate and precise records of people, traffic, military personnel or military vehicles. This vital information is recorded on a picture-map of the terrain under surveillance. Major reference points, such as a village, a railway line, a church, woods and hills, are first marked on the map and then the positions of the enemy units are superimposed on the drawing. Every few minutes the man on duty scans the countryside systematically from left to right, checking for any movement of man or machines. He then logs the time and movement. Today video cameras are sometimes used to provide the planners back at base with a real-life picture of the scene and the photographs are transmitted to HQ in digital form via a satellite communications link.

The expertise of the SAS behind enemy lines is often of the greatest value when friendly artillery and warplanes need to target enemy formations or military targets. This task is usually carried out by specially trained Forward Air Control teams, but the SAS is also quite capable of directing warplanes and artillery if necessary.

A little publicised area of the training undergone by all SAS soldiers is the methods that should be employed when on the run in enemy territory, when enemy forces are in hot pursuit, and when in a hostile environment. And each SAS unit under training is reminded that this part of their course might well be the difference between life and death.

The trainees are told that evasion is the best form of protection. Enemy ground and aerial patrols, some using infrared equipment, may well be searching for them and those on the ground will search the area for footprints, hides, old campfires, discarded equipment and any trace of recent human waste. When behind enemy lines, those trying to evade notice must take care never to be seen by anyone if at all possible, because that information can be passed to the searching troops within minutes.

The SAS recruit is taught that the evader must rely on his own eyes and ears as well as camouflage and fieldcraft. Naturally, recruits are told that it is a cardinal rule to move only during the hours of darkness and to hole up during daylight, making sure a new hide has been constructed before the dawn.

They are also taught how to live off the land while evading any human contact. They are taught what can be eaten safely and what to ignore. During the spring and summer in most countries there are edible plants, such as berries, some plants and the tops of ferns. On most continents there are also small edible animals, such as rabbits, hares, rats, mice, birds, fish and other small mammals, many of which are plentiful in broadleaf woodland.

The SAS survival course trains a man to make a range of animal traps and snares, but the instructors point out that luck is a very important element, no matter how carefully such devices are constructed, and in fact advise soldiers to concentrate more on catching fish. They have great fun in ordering trainees to catch snails, slugs and insects, and cook them, together with wild plants, in a large pot and then eat the succulent stew.

Men evading the enemy in farming areas are encouraged to eat a variety of root crops, and fowl (and their eggs), which are a great source of vitamins. But the instructors warn against killing and eating part of a sheep, lamb or goat, because that could be noticed by a sharp-eyed shepherd. These animals are also too large to cook and hide in secret. In any case, trainees are warned, it is unusual not to find dogs where there is livestock.

The most difficult and hazardous task facing all those evading capture is when the enemy is in hot pursuit and tracker dogs have been brought into an area where the evader is desperately trying to get back to his own lines. In such circumstances, SAS instructors are adamant, those on the run must eat raw food for fear that smoke might be seen. In addition, all rubbish and human waste must be buried deep enough to leave no trace for tracker dogs. It is better still, wherever possible, to put waste into a plastic bag, weight it down with stones and place it gently in water so that it sinks out of sight, leaving no smell for the dogs.

It is vital for a man who is being hunted to take meticulous care to leave no clues whatsoever. Throughout the hours of daylight he must lie still under a groundsheet, moving as little as possible and, importantly, should breathe down into the ground rather than into the air, so that the earth absorbs his body odour.

During the night, when he will be on the move, he should keep to lanes and tracks used frequently by the local people to confuse the tracker dogs, as well as taking every opportunity to walk along streams and any other water obstacle, so as to throw them off the scent.

If a group of men are trying to escape, they will find it easier to confuse the following tracker dogs by repeatedly taking off in different directions, coming together again and then moving away, criss-crossing one another's tracks. This not only confuses the dogs but makes the handler believe that his dog has lost the scent.

SAS men trying to make their way back from hostile to friendly territory are also taught the best way to behave and dress if it proves absolutely necessary to come out of hiding and mix unobtrusively with the local people. The advice is to walk confidently and obtain non-military clothing, such as overalls or jeans; remain clean-shaven; use bicycles and trains but keep away from railway stations; beware of dogs and children; and never walk along roads.

Trainees are told to assume that the enemy – both troops and police – will be actively searching the area for an escaper or evader. For this reason great caution must be taken in crossing roads, railways and rivers. They are advised never to cross a bridge, even at night, but to ford or swim a river, no matter how cold the water.

They are also taught how to improvise using everyday

materials. As every potential boy scout knows, fire lighting can be carried out with the help of a magnifying glass, waterproof matches, a flint stone or a cigarette lighter. Bootlaces can be used as fishing lines and needles and small pieces of wire as hooks or snares. Water can be carried in condoms or a plastic bag. Animals and birds can be skinned with any knife or even a wire-saw. Discarded tin cans make satisfactory mugs or mess tins for cooking. A sock filled with sand makes a good water filter; a small, hollowed-out piece of wood makes a spoon; old sacking makes a windproof coat; and string is a substitute for a belt.

And there are more potentially life-saving tips. Trainees are told about the best places to hole up in during daytime without expending effort or time constructing a proper hide. This ploy also has the advantage of making detection of the escaper's presence in an area more difficult. To keep dry and relatively warm, a soldier can bed down in a bramble bush, a hedgerow, a small cave or a cleft in rocks, or a hollow log. He can also nestle down in the shelter of a dry-stone wall, or in an animal burrow, a dry stream bed or a hollow tree. And dry leaves packed around him will not only act as camouflage but also keep him warm.

During training, SAS recruits are taken blindfold on a long journey in the back of an army truck and deposited in the middle of nowhere wearing army fatigues and supplied with a sparse amount of food and water. The trainee is given no map or compass, and must learn to navigate by night using the stars or the moon. Navigation by these means is difficult to learn and is not exact, but it does work.

Throughout their years in Northern Ireland the SAS became the thorn in the side of the Provisional IRA. Political apologists for the Provos tried their damnedest over many years to try to blacken the name of 22 SAS because it always beat the Provos at their own game. When the Provos took over from the less violent Official IRA after a vicious turf war in the early 1970s, the SAS were secretly dispatched to Northern Ireland to take out the Provo gunmen and bombers.

Because of the amateurish manner in which the Provos then planned and executed shootings and bombings, the SAS had no problem in picking up and dispatching a number of young Provos whose imagination and fervour had been filled with the songs and tales of heroic Irishmen who had died in a vain bid to kick the British out of Ireland. The Provo leadership concentrated on this kind of traditional propaganda in sing-songs in pubs and clubs where the cap would be sent round for funds for the boys in the front line. As a result, scores of young Irishmen were only too happy to join the ranks of the Provos to continue 'the fight for freedom'.

But the SAS stood in the way of the gunmen and the bombers. Though the SAS were quite capable of taking out Provo terrorists in the fields and byways of Northern Ireland, senior SAS officers believed that Special Force soldiers should not be used for patrolling high streets and main roads.

As Robin Neillands writes in his book *In the Combat Zone*:

As far as the Army and the SAS in particular are concerned, Northern Ireland has been a mixed blessing. Patrolling with 'a bullet up the spout' and learning to soldier when your life is at risk, however remote that risk, is a good way to give an army that extra little edge, but actions 'in aid of the civil power' should be undertaken only when all else has failed and will always carry the risk that soldiers will become involved in actions for which their training and their temperament has left them less than well equipped.

Most of the soldiers, even those in SF [Special Forces] and SOF [Special Operations Forces] units, are quite young and of junior rank, for terrorist wars are 'corporals' wars', and these young soldiers have to take on tasks that might be, or should really be, the responsibility of the police. One lesson that comes out of the

Northern Ireland business is that soldiers should be committed to counter-terrorism tasks in an urban environment only in certain circumstances and as a last resort – a lesson underlined by what happened when an SAS unit went into action on the Rock of Gibraltar in March 1988.

Ironically, in the light of this observation, the killing of three Provos – Sean Savage, Daniel McCann and a young woman named Maraid Farrell – was a classic Special Forces operation from beginning to end. Not only was the shooting of the three terrorists carried out by an SAS brick, but other SAS bricks were directly involved in related surveillance, guard duty and undercover recces over a period of five months.

Another SAS brick, working in plain clothes and under cover, was tasked with watching a well-known building contractor, Eric Martin, the Managing Director of H. & J. Martin, a firm which carried out many building projects for the security forces all over Northern Ireland. Building contractors and their workers – including secretaries, bricklayers and plumbers – had always been a principal target of IRA killer squads. Indeed more than a hundred innocent building workers had been killed by IRA gunmen during the first fifteen years of the troubles.

Eric Martin was a prime target because he employed some two hundred men from all over Northern Ireland and the IRA leadership believed if they could execute him, then the firm would fold and the Security Forces would be severely disadvantaged.

The Tasking and Co-ordination Group, who met on a daily basis, comprised senior Intelligence co-ordinators from MI5 and MI6, Special Branch, Army Intelligence, the RUC and the SAS. They decreed that a high-priority twenty-four-hour guard should be kept on Martin, and this task was given to the SAS.

The soldiers were housed in the double garage of Martin's home near Belfast, living there day and night and taking it in turn to stand guard during the hours of darkness. They never went near the house, some twenty yards away, but ate, cooked, slept and washed in the garage while keeping a watch over the Martin household. At night they also patrolled the grounds. Every day Martin was taken to and from work by armed members of the Headquarters Mobile Support Unit (HMSU) of the RUC.

Four weeks after the security net was thrown around Martin, HMSU officers noticed a young man acting suspiciously outside the premises of H. & J. Martin in Belfast's Ormeau Road. Photographs were secretly taken of the man and he was identified as Daniel McCann, aged thirty, a butcher from the city's Lower Falls area who had been jailed in 1980 for possession of explosives. In 1986 McCann had been promoted to IRA commander for the Clonard district of West Belfast.

From that moment a 'Det' team – a four-man SAS unit on Detachment operating in plain clothes – was tasked with watching McCann twenty-four hours a day, seven days a week. In the following days McCann was sighted in and around the lane leading to Martin's house and making a number of visits to the Ormeau Road. The Det team were certain that he was planning a raid for the Provo Active Service Unit (ASU).

As the days became weeks the team noticed that an attractive young, dark-haired woman would frequently accompany McCann during his recces and, from photographs, discovered that she was Maraid Farrell, a first-year student in the Faculty of Economic and Social Sciences at Queen's University, Belfast, who lived at home with her parents. One of her subjects was political science.

Independently, but around the same time, Gibraltar police became aware of a young, dark-haired woman who was paying frequent visits to the Rock from Spain, walking through the town and the area around Government House for an hour or so before driving back to Spain. Photographs were taken of her, sent to MI6 in London and passed to Northern Ireland. The woman was Maraid Farrell.

And in November 1987 a Spanish undercover team at Madrid airport spotted and photographed two suspicious-looking young Irishmen arriving from Málaga, in southern Spain. The two men were Daniel McCann and Sean Savage, another known Provo gunman, travelling under false identities. SAS bricks continued to watch over McCann, Savage and Farrell, who went to Gibraltar on two further occasions, in January and February 1988.

The IRA visits to Gibraltar were discussed at 10 Downing Street by the Central Intelligence Committee, the government's most secret committee, comprising the Prime Minister, Margaret Thatcher, the chief officers of MI5 and MI6, and members of Special Branch, Northern Ireland's Special Branch, and Army, Navy and Air Force Intelligence.

Britain had been constrained from arresting or killing any members of IRA ASUs operating in mainland Europe because International Law had to be respected. But Gibraltar was a different matter. Its defence, internal security and foreign affairs are controlled by the British government-appointed Governor. His authority was above that of the local legislature, the House of Assembly, which had limited powers. With the Governor's permission, British forces could operate freely on the Rock.

At the meeting it was decided to keep the three IRA suspects under constant surveillance and to have a unit from the SAS's Special Projects Squadron on permanent standby, ready to fly to Gibraltar at an hour's notice. A MI5 close surveillance team was sent to Gibraltar to draw up a list of possible IRA targets. They concluded that an IRA ASU would probably try to explode a powerful bomb somewhere on the Rock, but most likely at Government House.

On March 4 Maraid Farrell arrived in Málaga to join McCann and Savage. From that moment the three were kept under constant surveillance by the SAS and MI5 surveillance units. MI5 were convinced that the three would adopt the IRA's tried-and-tested method of using two inconspicuous cars. One clean car, devoid of explosives, would enter the Rock and park somewhere near the target and the 'bomb car' would arrive just before zero hour and take the parking place of the clean vehicle.

McCann was not spotted as he drove a car from Spain on to Gibraltar and, again undetected, he parked the ubiquitous white Renault 5 in the plaza. The SAS and the Security Services were very nervous. They had lost track of the three Provos and had found no bomb car. And the clock was ticking towards 4 pm, when the weekly ceremony of Changing the Guard, attended by hundreds of visitors, was due to take place. And this was the target which MI5 were convinced the IRA ASU planned to bomb.

At 2 pm two SAS bricks were patrolling the streets of Gibraltar in plain clothes and armed with Heckler & Koch sub-machine guns. They had seen many photographs of the three IRA terrorists and knew exactly who they were searching for. But they didn't know their whereabouts or the whereabouts of the bomb car.

Out of the blue came the vital breakthrough. Savage had been sighted fiddling with something inside a white Renault 5. The assumption was that he was priming the explosive charge ready for detonation by a radio signal. Minutes later the HQ that had been set up in the Gibraltar police chief's headquarters was informed that McCann and Farrell had been spotted crossing from Spain into Gibraltar. After Savage had left the car, an explosives expert quickly examined it and reported that it probably contained a bomb.

A radio message was flashed to one of the SAS bricks to tell them what had happened. They were ordered to take out the three Provos before they could explode the car bomb. McCann, Savage and Farrell were being tracked by four armed SAS men in civilian clothes who were only some twenty yards behind them. When Savage suddenly left the others and headed towards the town, two SAS men followed him while the other two continued to track McCann and Farrell.

Then the noise of a police siren shattered the peace and quiet of that sunny Sunday afternoon as crowds were gathering to watch the Changing of the Guard. McCann looked around him, wondering if the siren had anything to do with their bomb plot, and caught the eye of one of the SAS men some thirty feet behind him. The terrorist realised that the man was after him, and the SAS soldier knew that he had been spotted. The months of trailing and surveillance were over. It was the moment of action.

According to the evidence given to the inquest by the four SAS men, McCann was shot dead only after he put his hand inside his jacket – as if to draw a gun or detonate a bomb. Farrell was shot dead after attempting to open her handbag and Savage was shot dead only after putting his hand in his pocket.

But, on the day after the shooting, the government made a statement to the House of Commons revealing that none of the three Provos was armed, none of them was carrying any triggering device and there was no bomb in the Renault 5. Allegedly, a car packed with Semtex was discovered in a hotel car park in Marbella the following day.

As a result, controversy over the shootings went on for months. The main point of argument was whether the three Provos should have been shot dead rather than simply arrested and detained for questioning. In reality, was the killing simply the SAS putting into practice the Thatcher government's alleged shoot-to-kill policy? There was another problem. The SAS is, quite rightly, seen as the protector of the moral high ground and will act only within the law. There were those who suggested that the SAS could have very easily arrested the three without firing any shots.

But those SAS men didn't know the Provos weren't armed. They had been trained to shoot first and ask questions later rather than put at risk their own lives, and maybe those of innocent people, by seeking to ask questions first.

Throughout the raging controversy there was one major problem for the SAS soldiers which has never been adequately answered. As the three Provos had no guns and no triggering devices, why would any of them reach into their pockets or handbag? Evidence was also given at the inquest that a witness had seen the shooting from her apartment and said that both McCann and Farrell had both raised their arms as if surrendering seconds before the SAS soldiers opened fire. As a result, most people came to the conclusion that the four SAS men were under orders to shoot to kill, whatever the circumstances.

The shootings left many Members of Parliament uncomfortable with the government's policy of using the SAS in a plain-clothes, counter-terrorist role. Most would have preferred such incidents to be carried out by specially trained police units rather than the SAS. As a result of the Gibraltar killings, the SAS has been keen to pull back from its role as an arm of the political executive, preferring to leave such work to the police. Nevertheless, it does want to retain its involvement in counter-terrorism, which it sees as a most important element of its military duty, but it does not want to become embroiled in political arenas which could see its soldiers being used as hit squads.

Today all Special Forces in the Western world tread a very narrow path. In carrying out their duty to uphold democratic principles in support of an elected government, especially in the fields of surveillance, intelligence-gathering and ambushes, they risk usurping police roles and using excessive military force, which could well bring them into disrepute in the eyes of politicians and the public.

CHAPTER 8

US SPECIAL FORCES

O N OCCASION THE United States Special Forces work together on major operations, including counter-terrorism missions. Some of these are a great success, but others end in death and disaster despite the high calibre and commitment of the personnel involved.

One such hellish experience was the attempt by US Special Forces to rescue sixty-three Americans taken hostage during the invasion and occupation of the United States Embassy in Tehran by an army of some five hundred Revolutionary Guards in November 1979. Following the overthrow of the Shah of Iran, Reza Pahlavi, at the beginning of the year, Ayatollah Khomeini had returned to the Iranian capital from exile in Paris. The Islamic revolution had arrived in Iran and it would sweep the country, winning the support of both the poor and the middle classes.

One of the main targets of the young firebrand revolutionaries was the United States, which they dubbed 'the Great Satan'. Demonstrations, including burning of the Stars and Stripes, were almost a daily occurrence outside the US Embassy in Tehran and then, some ten months later and without warning, the demonstrators invaded the building and took the all of the staff hostage.

The United States tried to negotiate but was rebuffed at every turn. Other governments, including Britain and France, tried to intervene through diplomatic channels in an effort to persuade Ayatollah Khomeini to release the innocent hostages. All to no avail. The US President, Jimmy Carter, ordered the Pentagon to prepare a rescue mission.

The task was handed to Delta Force, a counter-terrorist unit which had only come into existence in 1977, after Colonel Charles Beckwith, who had served with Britain's SAS in the 1960s, persuaded the senior generals of the United States Army to form the 1st Special Forces Operational Detachment. Beckwith believed that the United States desperately needed an SAS-style Special Force dedicated to counter-terrorism and unconventional warfare. He was right.

Beckwith was given command of Delta and a training base at Fort Bragg, in North Carolina, and set about recruiting volunteers to join his new elite force. He adopted the same stringent tests as the SAS for those seeking to join the new force, as well as the same tough training schedules. Six months later Beckwith had selected just seventy-five men who he believed had the guts, determination and ability to make the grade. He was given two years to have two Delta squadrons up and running efficiently as a serious counter-terrorism force, ready for operations overseas. Incredibly, exactly twenty-four hours after Beckwith officially informed the US Army that Delta was operational, the Iranian revolutionaries invaded the US Embassy in Tehran and took the sixty-three hostages.

Beckwith and Delta were handed the tough task of rescuing the hostages. It would prove to be a baptism of fire for the fledgling Special Forces unit.

One week later Delta Force received orders to move to Camp Peary, in Virginia, the CIA's secluded base for training members of its Directorate of Operations in covert operations and counter-terrorism. Major General James Vaught, who had served with the US Rangers, was given overall command of the rescue operation, code-named Eagle Claw. Firstly, a detailed model of the twenty-five-acre Tehran Embassy compound was constructed. Measuring some eight feet by twelve, this showed the fourteen buildings contained within the wooded compound, which was protected by high walls. The model proved extremely useful in drawing up detailed plans of the rescue, and then thrashing out the optimum tactics and techniques, including the methods of entry and escape.

One month later a plan had been devised. A Delta Force unit of one hundred and twenty men, including a detachment of US Air Force ground crew, would fly in two giant Lockheed C-141 Starlifter transport planes to Frankfurt in Germany, where a thirteen-man US Special Forces unit which had trained to carry out the actual rescue of the hostages would join the flight. The C-141s would later fly to the island of Masirah, off the coast of Oman, from where the rescue plan would be launched.

The rescue unit would fly to a location code-named Desert One, in the Dasht-e-Kavir Salt Desert, two hundred and fifty miles south-east of Tehran, where they would be joined by eight Sea Stallion helicopters from the aircraft carrier USS *Nimitz*, which was then slowly cruising in the Gulf of Oman. The helicopters would fly the one hundred and twenty Delta Force soldiers from Desert One to some fifty miles south of Tehran.

Great care had been taken to ensure that the whole operation would go smoothly. Those in command knew that the eyes of the entire world would be on this mission, for hardly a day passed without TV, radio and newspapers reporting whatever details they could discover about it. On the front of some newspapers and in numerous news bulletins a figure would appear showing how many days the hostages had been in captivity. And throughout the United States yellow ribbons began to adorn trees – a powerful symbol of the importance the men, women and children of America placed on the insult to their country perpetrated by the Iranian hostage-takers. Hardly a day passed without appeals being made, plans drawn up, ideas suggested of how the United States could bring this tense situation to an end.

Two weeks before the Delta Force unit flew to Frankfurt a four-man recce unit was secretly smuggled into Tehran to find suitable landing sites for the choppers, routes for the entrance of the Delta soldiers and exits for them and the hostages, some of whom were likely to need physical assistance.

The plan was for the Delta soldiers to split into three groups. Red Group would attack the western side of the embassy compound and release any of the hostages it discovered. Blue Group would take the eastern side. White Group would secure the withdrawal route for those hostages brought out by Blue and Red Groups as they were ushered to a sports stadium opposite the embassy. From here the hostages would be airlifted by waiting US helicopters to a location some thirty miles from Tehran, where US aircraft would be ready to fly them out of Iran.

Throughout the rescue operation two AC-130 Spectre helicopter gunships would orbit the compound, suppressing any opposition from the Iranians guarding it and the hostages.

On Sunday April 20 1980 the Delta unit flew to Frankfurt and on to Egypt, where their weapons and equipment were given their final check. Four days later they flew on to Oman. As dusk fell that night the men took off in three smaller MC-130 transport planes for Iran, as did the eight Sea Stallions from the *Nimitz*.

However, earlier in the day bad luck had dogged the operation from the very start. A road ran alongside the landing site, and no sooner had the eight-man team deployed to watch the road for any danger than a bus packed with passengers trundled along it towards them. The passengers had a wonderful view of three US aircraft standing idly in a deserted area. The US unit opened fire, aiming to bring the bus to a halt by shooting out its tyres. It worked perfectly; the bus stopped and the Delta team ushered the frightened and surprised passengers off the bus and ordered them to sit on the ground and wait. Four men were left to guard them.

The passengers had just been herded into the field when a petrol bowser came down the road. One soldier fired a single M72 light-armour shot, hitting the bowser, setting it on fire and sending a great ball of flame shooting up into the night sky. It could be seen for miles. Within minutes a light van came down the road, picked up the bowser's driver and mate and disappeared into the night, never to be seen again. That was serious.

However, the delivery of the full complement of one hundred and thirty-three troops in Iran went smoothly, with the US transport planes depositing the men and their equipment and taking off again for Oman. But the eight Sea Stallions, which were to fly the troops to within fifty miles of the embassy, were very late. Mechanical breakdown had forced one to be abandoned and instrument failure had forced another to return to the *Nimitz*. The other six finally made it to the pick-up point, but they had run into severe sandstorms which had reduced visibility to zero.

As the Delta soldiers were clambering into the six helicopters, one of them sprung a leak in its hydraulic system and had to be abandoned. Now the rescue force was down to five instead of the planned eight choppers. The highly experienced Colonel Beckwith, who had been responsible, with other senior Special Force officers, for planning the entire rescue operation, knew it was extremely dangerous to continue with only five helicopters. The decision to 'stop or go' went to President Carter himself, who ordered the mission be aborted. Beckwith gave the order for the entire force to fly out on a single MC-130 transport plane and three EC-130 helicopters and for the remaining Sea Stallions to be destroyed.

It was at that moment that disaster struck. As Peter Harclerode wrote in *Secret Soldiers*:

As one of the helicopters was manoeuvring to refuel from one of the EC-130s at the northern end of Desert One, its rotors struck the port side of the tanker which exploded in a fireball. Members of the Blue Group already aboard the EC-130 were forced to leap for their lives as fire engulfed both aircraft, killing the EC-130's five crew and three others aboard the helicopter. Total confusion now reigned throughout Desert One as the flames turned night into day and exploding ammunition and Redeye missiles provided an impromptu fireworks display. Without further delay, all personnel boarded the three remaining aircraft which took off for Oman. In the haste to depart, however, the destruction of the five remaining, now abandoned, helicopters appeared to have been forgotten.

Operation Eagle Claw had been an unmitigated disaster, highlighted by the loss of eight lives. The abortive raid, the first ever undertaken by Delta Force, was hugely embarrassing to the US Special Forces, and Delta in particular, the US Marine crews who had abandoned the helicopters, the planners and President Jimmy Carter and the United States Army. Indeed across the world the mismanagement and apparently amateur approach to the rescue mission showed the US armed forces in a dreadful light. They became the butt of jokes the world over, particularly among the armed forces. Worse still, they eventually became the target of jokes on TV. It was a ghastly, humiliating episode.

Later it was revealed that not only had the US Marine crews abandoned the five helicopters in a totally unprofessional decision, but, unbelievably, they had even left intact on board the choppers' secure communications equipment and documents giving details of Operation Eagle Claw and of US agents within Iran. As a result, some of those US agents were picked up by Ayatollah Khomeini's revolutionary guards and executed after suffering the most horrendous torture.

This debacle, however, has been used ever since in Special Forces training lessons throughout the world. There was nothing intrinsically wrong with the planning, the training of those taking part or the commitment of the Delta Force, the US Marine Corps, the US Air Force or other members of the US armed forces who were involved in the operation. What went wrong concerned the professionalism and the attitude of some of those who took part. For example, it seemed that the Marines had totally failed at the moment of truth, not even caring what highly secret documents they were leaving behind for the enemy to peruse, not bothering to destroy the helicopters they had been forced to abandon. Such measures were a matter of common sense and yet, when the heat had been turned up and people had to think on their feet, they simply weren't up to the mark.

It is noteworthy that after the failure of the US Marine Corps in that ill-fated venture it was decided by Pentagon chiefs that Delta Force needed its own helicopter unit in the same way that the SAS has its own dedicated chopper and transport unit on permanent standby. In 1997 Delta's order of battle was upgraded and now includes an Aviation Platoon with its own fleet of helicopters.

When President Ronald Reagan took office in 1981 he let it be known to friend and foe alike that he was determined to ensure that the United States would not be pushed around by anyone again, and would not permit any nation to cock a snook at Uncle Sam. Furthermore, he was determined that the nations of Latin America and the Caribbean, some of which were then flexing their muscles and opting to follow Cuba's example, would not be permitted to seize power and set up hard-left or communist regimes in America's backyard.

On October 12 1983 the elected government of the tiny island of Grenada – a left-wing junta led by Maurice Bishop – was overthrown in a coup led by the hard-left New Jewel Movement. One week later the popular Bishop and six other government ministers were taken from their prison cells, lined up along the walls of Fort Rupert and shot dead in a traditional Latin-American execution. Bishop's execution brought demonstrators on to the streets of Grenada in protest, but the island's new leaders, also left-wing, replied with bullets. The situation was deteriorating rapidly.

Bishop had been put into power with the help of Cuba's Fidel Castro, and Soviet government finance had been used to construct a ten thousand-foot runway capable of handling long-range transport planes, a new control tower and sophisticated airfield facilities. Allegedly, all this was intended to encourage tourism. But the CIA believed the runway would be used as a first-rate facility for Soviet transport planes to stop over when bringing material, weapons, ammunition and supplies to support popular revolutionary movements in Central and South America.

Studying medicine on the beautiful island of Grenada at that time were six hundred American students, and the Reagan administration was worried that here was a hostage crisis just waiting to happen. The young Americans could easily be taken as pawns in a dangerous, high-stakes ploy aimed at stopping the US armed forces from invading the island. But in fact the small nations of the region looked to the United States for leadership because they feared that a powerful left-wing Grenada, backed by the Soviet Union and Cuba, might be a threat to the entire region.

Reagan decided to launch Operation Urgent Fury. It was only four years since the debacle in Iran, and Reagan and the Pentagon were determined that this operation would go like clockwork, whatever the cost and whatever the number of troops needed to ensure total success. Reagan was also keen to show himself as a strong leader, a president who was prepared to use force once again whenever the need arose. He also wanted to overcome the nation's defeatist attitude following its humiliating and costly withdrawal from Vietnam. He told the Pentagon to throw everything into the operation and to make sure that it was totally successful.

Grenada, only some fifteen miles long and ten miles wide, was at one time a part of the British Empire but became an associated state within the Commonwealth in 1967. In the 1980s its army, the Grenadian Defence Force, totalled about three thousand soldiers armed with only rifles, pistols and sub-machine guns. They were backed by some seven hundred and fifty Cuban troops, euphemistically described as 'construction workers'.

Against this enemy the United States launched an attack which included the 22 Marine Amphibious Unit, a team of Navy SEALs, elements of the 82nd Airborne Division, a small fleet of warships and, from the United States Air Force, one hundred and sixty Black Hawk helicopters, a Tactical Fighter Wing and the First Special Operations Air Wing.

At dawn on October 25 1983 the invasion was launched when Delta Force soldiers were ferried by air to an airfield at Point Salines, which they were tasked to seize and control to enable USAF transports to fly in with troops and supplies. As the Delta teams arrived, two battalions of US Rangers – some sixteen hundred men – parachuted on to the airfield. The Cuban defenders seemed totally overawed by this dramatic illustration of US combat power. There were a few brief firefights with defenders, but within thirty minutes the Delta teams and the Rangers had full control of the airfield, the all-important control tower and all the other buildings.

But the battle for Grenada was by no means over.

The United States was keen to ensure the safety of other Grenadian politicians, dignitaries and others who, it had been led to understand, had been rounded up and incarcerated in Richmond Hill Prison. It was fearful that the new government might decide to slaughter all the 'political' prisoners in retaliation for the US invasion in the same way that they had executed

Maurice Bishop and his ministers without trial. The attack on the prison was undertaken by Rangers in six Black Hawks. But the intelligence passed to the US Special Forces units taking part in the invasion proved to be poor and totally inadequate. The CIA had once again failed the US military and it would be members of the courageous Special Forces units who would suffer and die as a result. As they attacked, the Black Hawks found themselves under fire from the latest, most advanced and sophisticated Soviet-built ZSU-23-2 (twin 23mm) anti-aircraft guns. The United States had no idea the Soviet Union had installed these in Cuba, let alone tiny Grenada.

Then, as the helicopters circled to land, the Soviet anti-aircraft guns on Fort Frederick, just three hundred yards from the prison, opened up, sending two of them spiralling out of control to the ground and killing the crew and eight Rangers on board. The other four Black Hawks were able to take evasive action, but they could find nowhere to land. Two hovered low enough for SEALs to fast-rope down and rescue their wounded comrades just as the People's Republican Armed Forces arrived on the scene. A firefight developed within seconds, with the SEALs desperately trying to protect their wounded comrades and keep the PRAF troops at bay. The SEALs called up another helicopter gunship, which arrived an hour later, just as their ammunition was running low. The gunship wreaked havoc on the enemy forces, killing and wounded some and scattering the rest. It was not until the following day that the SEALs and their wounded comrades could be pulled out by US helicopters and flown to safety. Altogether twelve Rangers were wounded in an attack that had been a total disaster.

On the southern shores of Grenada, Delta paratroopers also had a tough time as the wind pushed them out to sea. They had been dropped some twenty miles off the coast, along with two inflatable craft. Waiting below were two small whalers, manned by US Navy forces, which would take them to the island. But the wind and heavy seas made it very difficult for the boats to locate and pick up the SEALs. Some twenty Delta men had major problems in the rough waters, battling against the current to the shoreline and weighed down with all their equipment. Four drowned as they tried to link up with the surface vessels.

As a result, one whaler stayed offshore searching for the lost SEALs while the other continued towards the shore. But this second boat had to heave to and lost all power, its engine soaked by the rough seas. The mission was aborted and they tried exactly the same battle plan the following night. That also ended in failure when the two boats were again subjected to a heavy battering which drenched the SEALs. It was decided the invasion of Grenada would continue but with no forces being put on the southern part of the island.

There was another objective, also given top priority by the US administration – the safety of the Governor-General, Sir Paul Scoon, the British government's representative, who was still living in Government House. A SEAL team of twenty-two men was flown in seven Black Hawks straight to Government House and, despite anti-aircraft fire from nearby Fort Rupert, within minutes some men had fast-roped on to the roof of the building while the rest had landed in the grounds, taking the Grenadian defenders totally by surprise. The SEALs in Government House took up defensive positions and shepherded Sir Paul and his staff into the ground-floor dining room.

But the PRAF soldiers on guard duty that night were alert to the situation. They immediately called up their defence forces and two Soviet-built BTR-60 APCs with multiple medium machine guns trundled into the grounds and opened fire on the SEALs as they were organising their battle plan. It had been intended that the SEALs would evacuate everyone from Government House, put them into helicopters and fly them to safety. Now everything had to change, for these Soviet APCs were causing mayhem among the SEAL team, who were equipped with no more than rifles, sub-machine guns and pistols.

The SEALs had no protective armour, no defence and pathetic weapons. There had been no suggestion in any pre-attack briefing of the possible presence of these powerful Soviet APCs, and this annoyed the SEALs officers on the ground. Once again the bravest of US Special Forces had been let down by appalling lack of professionalism by the CIA and US military intelligence. The SEALs in the grounds, most hiding behind trees, kept up a steady fire, but their rounds were making little or no impact on the APCs and they were running out of ammunition.

But one of the US helicopters made contact with US Air Command controllers, detailing the gravity of the situation on the ground and calling urgently for helicopter gunships to be sent immediately to save the lives of the SEAL team pinned down on the ground and with no chance of escape. The two APCs kept reloading and raking the ground with withering automatic fire. Every time any of the SEALs tried to make a dash to a different area in a bid to outflank and get behind the APCs, the machine guns would open fire, often hitting the courageous soldier. Those SEALs in Government House kept up continuous fire on the APCs and prevented any enemy ground forces entering the compound.

An hour later a Spectre finally arrived above Government House, but this was only just sufficient to carry out the job. The helicopter gunship stood off and blasted the two APCs with everything, raining missiles and bullets down on them until they were destroyed. Even so, they had managed to continue firing for a further twenty minutes. But there was still danger for the SEALs in the compound. Nine of them were lying on the ground seriously wounded and in urgent need of medical care.

Another Spectre was called in and this succeeded in driving most of the enemy forces away from the immediate vicinity of the compound. The SEALs managed to retrieve their wounded comrades and move them to the safety of Government House. For the rest of that day and night SEALs on guard duty managed to keep any enemy forces at bay until the following morning, when US Marines arrived.

And that was not the end of the matter. The PRAF proved a tough nut to crack despite the US forces' massively superior firepower. Pockets of resistance sprang up all over the island and it was not until a week later that the American forces could announce that the island had been completely overrun and hostilities were at an end. In the final tally some one hundred Grenadian and Cuban soldiers were killed and over three hundred had been wounded. US casualties totalled eighteen dead and fifty-seven wounded. This victory had been no easy pushover.

But the invasion of Grenada was hailed as a success and the United States military would claim that it had proved it could organise a rapid response force in an emergency and succeed. More importantly, the US administration had shown the United States could look after affairs in its own backyard of Central America and the Caribbean. Some weeks after the invasion it decided to take no further chances with any other possible Marxist-led uprising in Central America or the Caribbean islands and one hundred Special Forces teams, each consisting of eight Green Berets, were deployed in Honduras, Costa Rica and Guatemala, as well as the islands of Grenada, St Kitts, St Lucia, Barbuda and Nevis, to train the local police and defence forces in counter-terrorist warfare.

During the past twenty years a number of top-secret missions have been carried out by US Special Forces, some of which have never before been disclosed or admitted by the US government.

One of these raids into foreign lands was ordered by President Reagan, on the advice of the CIA, after the United States decided to take drastic action against certain countries, such as Libya, which it was convinced were encouraging, supporting and financing terrorist action against the world's democracies. The Libyan leader, Colonel Gaddafi, had been using his wealth from oil exports to train Palestinian and other Arab fighters in camps in Libya, providing a safe haven for them to live in and encouraging terrorism across Europe – all in the name of Islam.

Gaddafi believed he was untouchable, constantly moving his headquarters to different parts of Libya in order to keep one step ahead of western Special Forces, such as the SAS and the SEALs, who he was convinced were determined to assassinate him. He wasn't far wrong. Indeed President Reagan launched a famous bombing mission against Gaddafi in the 1980s, destroying his famed tented headquarters in the desert, a place where he believed he was safe from attack. An unspecified number of Libyans were killed and wounded in the heavy raid, carried out by US warplanes, but Gaddafi managed to escape. The intelligence provided as to his whereabouts that night had been pinpoint accurate, but luck was with him and the bombs, for although they hit their target, they didn't manage to kill the man who was creating terror and murder in many European countries.

The United States was keen to show Gaddafi that if it really wanted to assassinate him it was perfectly capable of doing so. It wanted to warn him, to fire a shot across his bows, in the hope that he would end his support for Islamic terrorists. After the raid on his desert headquarters the Americans had also warned him that, if he did not end his support for terrorism, bombing raids might continue from time to time.

In June 1984 units of the French Foreign Legion had secretly taken a sweep through southern Libya from Chad, a friendly neighbour opposed to Gaddafi's support for international revolutionary terrorist groups, after being informed by the United States that the Libyans had set up a training camp for

Palestinian and European terrorist groups. United States spy satellites had discovered the new camp from photographs of Libya taken from space. Foreign Legion troops spent some weeks under cover in Libya and finally reported back that they had found the new camp – in the middle of the desert. Their recce showed that Libyans were training several hundred young terrorists, not only from Middle Eastern countries but also from Germany, France, Spain, Italy and Northern Ireland. The Foreign Legion returned with detailed information about the camp and vital film footage. The film, which had been obtained at great risk, showed not only the layout of the camp but also the training, the weapons, the instructors and some of the young men undergoing training.

The United States needed no further proof but it was decided that Special Forces should be employed to carry out a secret and devastating raid against the training camp rather than undertake a far more public bombing raid. This would be an attack tailor-made for a Special Forces unit working in secret.

In July, after weeks of detailed planning and training, a Delta Force unit of some sixty troops was flown to northern Chad in a USAF transport plane. They spent one week acclimatising to the scorching weather conditions in the African country, where temperatures by day were usually around one hundred and forty degrees Fahrenheit in the shade. This raid would give Delta Force an opportunity to live up to its famed initial 'D' for destruction.

Three Black Hawks, which had been transported to Chad in US C-141 transport planes, took off from the base camp in the early hours of the morning and flew low over the desert, arriving at the Libyan camp when the place was sleeping. As they approached the gates the attack helicopters flew at some twenty feet off the ground and sprayed the guardroom with a heavy machine gun, ripping the wooden building to pieces and killing those on duty. Two other Black Hawks arrived at the other side of the camp and some sixty Delta troops leapt from them and raced to their targets.

All those men taking part knew the layout of the camp by heart because a model had been constructed back in the United States to assist in the planning of the raid down to the smallest detail. It proved brilliantly successful.

One team ran pell-mell for the radio communications centre to ensure that no one would be able to send out a distress call saying the camp was under attack. When the six Delta men burst into the room, shooting out the door locks, the place was empty. They opened up with their sub-machine guns, destroying all the communications in the room. Another Delta team ran to the headquarters block, which was also empty. First, they ransacked drawers and cupboards, taking away with them any files that seemed, from quick scrutiny, as if they might contain valuable information. Then they threw grenades into the empty rooms, torching the place. The fires spread rapidly through the block until the entire building was in flames.

By now camp staff and the trainee terrorists were awake and grabbing for their weapons. But they were too late. Five Delta teams had run towards the dormitory blocks and covered all the entrances, while others had gone down the sides of the single-storey dormitory buildings, hurling grenades through every window. When anyone emerged they were met by a hail of automatic fire and killed instantly. The Delta boys were in no mood to deal lightly with these potential killers. Their orders were to wipe out everyone at the camp and ask no questions.

Only when the dormitory blocks were well alight and the flames were destroying the buildings did the Delta teams pull back, sure that no one inside was alive.

But the Delta teams hadn't yet finished the job. They had brought with them powerful demolition explosives, which they proceeded to place in the armoury and the magazine block.

The buildings went up with tremendous explosions that blew them to pieces. But the Delta men didn't bother to stop and check what was left. The time had come to get the hell out of the camp and out of Libya. It was undoubtedly a first-class raid by Delta Force, well planned, well executed and totally successful.

The world was to know nothing of this operation. President Reagan said nothing, nor did the US Army, Navy or Air Force. And Colonel Gaddafi said nothing either. Of course, he was seriously embarrassed by the raid, which showed that his military could not even defend their own camps against a quick-fire raid by American forces.

It seemed that the United States Special Forces had come of age.

CHAPTER 9

SCUD BUSTERS

T HE GULF WAR – code-named Operation Desert Storm – began on August 2 1990, when Saddam Hussein's forces invaded the oil-rich sheikhdom of Kuwait in the Persian Gulf without warning. Within hours the Iraqis had overrun the minuscule Kuwaiti defence force and taken Kuwait City. This occupation would become a perfect arena for Special Forces operations.

It soon became obvious that Saddam Hussein had no intention of pulling back his forces and indeed he set about pouring armour, aircraft, troops and infrastructure into the place quickly as possible. The United Nations Security Council ordered him to leave but he took not the slightest notice.

Fearful that Saddam Hussein's invasion of Kuwait could lead to his exercising a much tighter grip on the world's oil markets, the United States took the decision that the western nations must work together and kick him and his forces out of Kuwait – at whatever cost. Under the auspices of the UN, the United States, backed by the armed forces of Britain, France, Egypt and Saudi Arabia, took on the job of expelling the Iraqi forces. The principal reason for this unusual alliance was to show the world that, following the implosion of the Soviet Union and its

European satellite nations, there was to be a new world order, with the United States leading the way.

Saddam Hussein had poured one hundred thousand troops into Kuwait and some of those – particularly his elite Republican Guard – were believed to be a well-trained, well-disciplined, hard-fighting force which had performed with great credit during the Iran–Iraq war of the 1980s. The United Nations force, under US General Norman Schwarzkopf, believed that it would take weeks, possibly months, to drive the Iraqis out of Kuwait.

Schwarzkopf's army contained a surprisingly large element of Special Forces, including Britain's SAS and the SBS, and the United States Green Berets and Navy SEALs, as well as a number of Special Operations Forces, including the 1st and 2nd US Marine Divisions and the 101st Air Assault Division, this last unit being better known as the Screaming Eagles.

Schwarzkopf placed all the Special Forces contingents under one command – SOCCENT (Special Operations Command, Central Command), commanded by US Colonel Jesse Johnson. This Command was subdivided into three Commands – the Army Operations Task Force, the Navy Special Warfare Task Force and the Air Force Special Operations Command. These had some seven thousand troops under their command, including 150 SAS and SBS men, the 55th US Special Forces Group, SEAL teams 1 and 2, a detachment from the French Foreign Legion and some Special Forces units from Arab nations, including Kuwait.

From the outset Schwarzkopf and his senior commanders knew that Iraq possessed a number of Scud missiles capable of carrying a variety of warheads, including explosive, biological, chemical and nuclear devices. The Coalition forces were led to understand that Saddam Hussein did not have a nuclear capability, but the CIA were seriously concerned that he did possess both chemical and biological warheads. It was also accepted that Saddam Hussein would probably use these gruesome weapons if he felt it necessary. And it was the urgency of tracking down and destroying the Scuds that necessitated the use of such large numbers of Special Forces in Desert Storm.

It was further understood that Israel would be the likely target of such an attack, but this did help the Allies, and particularly the Special Forces, because they could concentrate on those areas of north-western Iraq from which the Scuds would have to be fired to reach a target in Israel.

The bad news was that these Scuds could be fired either from set positions, hidden in bunkers, or from the back of huge transporters which had been seen and identified by US satellite photography. These pictures had been taken during exercises in the desert by Iraqi Scud platoons. The photographs revealed that the transporters could only travel at slow speeds, but after reaching their destination the Scud platoons needed only an hour to prepare to fire their missiles. If Special Forces on the ground could identify the exact place where the Scud transporter had stopped to make ready to fire a missile, there would be enough time for US warplanes to scramble and strike at the missiles before they could be launched. However, the intelligence would need to be pinpoint accurate, and to be able to send back such precise grid references the Special Forces would have to take extraordinary risks.

On the first night of the war – Thursday, January 17 1991 – Saddam Hussein revealed his hand when eight Scuds carrying explosive warheads landed in Israel – two in Haifa and four in Tel Aviv. As well as also killing some one hundred innocent men, women and children and causing considerable damage this spread great consternation among the frightened population. Saddam Hussein's plan was to entice the Israelis into reacting against Iraq, which, he believed, would spark an Arab–Israeli conflict. He was convinced that, if Israel attacked Iraq, countries such as Jordan would take Iraq's side and countries such as Egypt and Saudi Arabia would be forced by public opinion on the Arab street to pull out of the UN Coalition.

This put Schwarzkopf under enormous pressure to take out the Scuds as speedily as possible for fear that Israel might decide to go it alone and send its war planes to bomb Baghdad. Indeed in some Coalition quarters it was feared that, if Iraq used chemical or biological Scud weapons, there was every chance that Israel would launch a nuclear attack on the Iraqi capital.

Therefore, when the Special Forces units were briefed by their commanders, they were left in little doubt that their missions were of the utmost importance.

The first Special Forces action of the war was Special Boat Service raid on January 23 1991, when two Chinook helicopters took twenty-four SBS Royal Marine Commandos to a grid reference near Highway 8, where the main Iraqi communications link between Baghdad and Kuwait was known to run. The Chinooks landed in the dark some two hundred yards from each other in case one accidentally ran into an Iraqi ground force. If such a force had closed on one Chinook there would still be the twelve SBS men in the other chopper to act as back-up. But they met no opposition.

The SBS came equipped with picks, shovels and cutting equipment. Some dug along the line they had been given, while the others spread out to keep watch. That night the SBS displayed textbook expertise. Wearing desert camouflage, their faces blackened, their bodies all but invisible as they lay along the highway route, the lookouts remained motionless as the others dug for and then severed the all-important line that would cut off Saddam Hussein from his front-line commanders in Kuwait. The mission went perfectly and within two hours all the SBS men withdraw to their Chinooks, taking with them a six-foot length of communication cable, which they later presented to General Schwarzkopf 'with the compliments of the Royal Marines'.

But although the SBS had carried out the first mission, the SAS had been out in the deserts of Kuwait and Iraq before the war had started. Shortly after the initial political decision was taken by the United States and Britain that Iraqi forces must be expelled from Kuwait, the SAS were on their way. By December 1990 the SAS had four squadrons – some two hundred and forty men – on the ground, ready for action. They had been sent to the Persian Gulf with specific orders to find the best ways of rescuing hostages seized by Saddam Hussein's forces. The Iraqi leader left the Coalition in no doubt that these European men, women and children were to be used as human shields. He ordered they be taken to potential Coalition targets such as power stations, telephone exchanges, command centres, army bases and armed forces headquarters. This policy of dispersal made it extremely difficult for Coalition bombers to strike at any Iraqi targets and for the SAS and other Special Forces to rescue the hostages.

While the SAS sat on their hands and tried to make feasible plans, diplomats were working behind the scenes to persuade Saddam Hussein to send home the hostages, for they were totally innocent people. And the Iraqi leader came to realise that by taking innocent hostages and using them as human shields he was setting the entire civilised world against his country. The policy worked, for he freed the hostages and permitted them to go home before the war even began. But it had been a close-run thing.

There were now two vital objectives for the SAS. The first and most important was to find and destroy the mobile Scuds and the second was to disrupt and cut the Iraqi main supply routes, attack convoys and destroy any depots or communication links they came across. This second task was a traditional SAS operation for which they had trained for most of their Special Forces career.

One squadron was divided into fighting mobile columns, which meant that all the men from Air, Boat and Mountain Troops had to re-train in vehicle skills, learning how to drive across deserts and also hone up on their shooting skills. All had to learn to fire Milan anti-tank missiles, 0.5 inch Browning heavy machine guns and American M19 40mm grenade-launchers. They practised for twelve days, during which they also had to learn how to ride motorbikes at speed across the desert, a demanding and physically exhausting task. A number also had to learn to drive long-wheelbase 110 Land Rovers which were equipped with medium machine guns and had all the kit and the mortars stacked on the back.

Each troop of eight Land Rovers took along a 'mother vehicle', a three-ton truck loaded with everything they might need, including spare ammunition, mortar shells, ammunition, spare rations and kit and the NBC kit – protective clothing used in the event of nuclear, biological or chemical attack. For greater safety, the mother vehicles stayed in the middle of the column of eight. Most of the time two motorbikes led the way, riding some two hundred metres in front of the lead vehicle, scanning the horizon.

Some SAS men complained that when they set out it seemed as though they were each carrying far too much gear to be truly mobile, and yet it was all very necessary kit, especially for the emergencies that would follow. Most carried an M16 assault rifle and a 9mm Browning high-powered pistol, sixteen magazines of thirty rounds each for the rifle and a couple of magazines of 9mm ammunition for the pistol, plus six grenades and a full evasion kit.

There was one major complaint from all the British Special Forces sent into the Iraqi deserts – clothing. All Special Forces had been advised to take light tropical clothing because the weather would be warm, even at night. It was an unbelievable basic error on the part of the planners. They should have known full well that at night during December, January and February the deserts of this region are always cold and often near to freezing. To send the Special Forces out dressed in warm-weather clothing was almost criminal. Some suffered frostbite, all felt bitterly cold and most were unable to sleep in the few hours possible because they were simply too cold. Indeed some of the men became so desperate that they pleaded with base camp to send warm clothing, any warm clothing, to keep out the cold. None had been sent to the Gulf by the Ministry of Defence, so quick-thinking quartermasters went to local markets and bought a load of thick, warm Bedouin coats worn by the shepherds and sent them by helicopter to prearranged meeting places. One SAS soldier remarked later, 'These coats were real lifesavers.'

One of the first SAS mobile units to cross into Iraq after the start of the war in search of Scuds ran into a large group of Iraqi forces which included ground troops and motorised patrols. Eight SAS men in two Land Rovers were cut off from the main column and surrounded. Under orders, the rest of the patrol drove on, leaving the eight men to fight their way out as best as they could. They were never seen or heard of again and no one ever discovered what happened to them.

The SAS travelled only at night and, thanks to their night-vision goggles, almost always saw the Iraqi forces before they were seen by them. In fact it seemed that the desert, especially near the main supply routes, was covered in enemy forces driving hither and thither, and on many occasions these clearly had no idea that the line of vehicles they could see on the horizon or even nearer was not friendly forces but an SAS mobile patrol going about its deadly business. Night vision also made it possible for the SAS patrols to approach remarkably close to the enemy forces before opening fire, because during the hours of darkness the Iraqis simply took them to be friendly forces.

One of many such contacts occurred when a unit of four heavily laden Iraqi trucks led by a jeep was driving towards an SAS mobile patrol of eight Land Rovers and its mother vehicle. The SAS men were ready to open fire, their heavy machine guns loaded and cocked for action as the Iraqis drew almost level. Suddenly there was a roar from the Iraqi officer in the lead jeep as he realised the vehicles coming towards him were not Iraqi. The SAS opened up, driving slowly past the convoy of four and pouring metal into the trucks before the enemy troops had time to open fire.

As soon as the last SAS vehicle had passed they all turned and came back. This time there was some rapid fire from the Iraqi trucks, but not machine-gun fire, and the SAS let fly once again, ripping the convoy to pieces. The jeep was destroyed, three trucks exploded and were set on fire by Milan anti-tank missiles and the fourth truck crashed into one of the others and burst into flames. The SAS didn't bother to stay and check if anyone was alive but drove off, changing direction after five miles in case someone had managed to give a grid reference back to base.

Occasionally the SAS mobile patrols would see some odd, interesting, possibly military, building on the horizon and would call base, giving grid references and air positions if possible.

One such discovery behind the enemy lines was a communications tower, an ideal target for an SAS mission. The tower, with its antennae, was some two hundred feet high and stood out like a lighthouse on the undulating desert. The SAS team waited until dark before getting a closer look because there was a lot of Iraqi traffic driving up and down the main road near to the tower. Under cover of darkness they moved closer and, through their thermal imagers, realised the tower was a major Iraqi communications centre. The tower was defended by a perimeter wall and fence and surrounding it were a number of small buildings, some temporary, and sixty or more military and civilian vehicles.

After spending several hours surveying and detailing the building, the defences, the guards and the overall layout, the SAS men withdrew a safe distance into the desert to plan their attack. Over their wireless link with base camp they requested any information about the tower and told of their decision to take out the place and cause as much damage as possible. Base camp confirmed that the communications centre was a vital link for the Iraqi military and that below ground there were believed to be three more floors stuffed with communications gear, forming one of the Iraqi Army's main communication hubs. It was not known at base camp how many guards were on duty or whether the camp housed a battalion or simply a small section of soldiers.

When the SAS group carried out their survey from the ground there had not appeared to be too many guards on duty at any one time. But they were surprised at the intelligence that there were three underground floors. It made them even more determined that this raid had to be both substantial and successful. They believed, as ever, that surprise would be the key. They only hoped that the guards were few and far between and lightly armed.

If the entire building with its vital communications equipment inside were destroyed, the SAS men believed, it would cause the Iraqi high command severe problems for some time to come.

The mother vehicle and the other Land Rovers were left behind, hidden from view in a wadi, or dry river bed, while three Land Rovers, with twelve men on board, moved off towards the objective in the early hours of the morning.

It was fortunate that there was no moon that night for the SAS were able to approach the tower without attracting any attention from the occasional Iraqi vehicle seen travelling along the main road some half-a-mile distant. It was a bumpy ride across the rugged desert, but the troops had no wish to take the main road and risk meeting an Iraqi army vehicle.

They intended to hit the tower just before dawn when they hoped the guards would be not at their most alert and all those inside would be fast asleep. A mile from the tower the Land

Rovers were parked out of sight, on the main road in a shallow dip in the terrain, and nine of the SAS men continued the rest of the journey on foot. They carried their Heckler & Koch machine-pistols and ammunition, the explosives, charges and timers, an anti-tank missile-launcher and missiles as well as grenades.

They moved slowly, weighed down with the excess baggage which they intended to leave behind after their attack on the tower. As they moved closer to the objective they had to keep freezing whenever Iraqi vehicles passed along the road running nearby the tower. One hundred yards from the tower three SAS men crawled forward on their bellies to lay the explosive charges which they hoped would blast open the wall sufficiently to give them direct access. The others stayed back with their anti-tank missiles at the ready in case they were needed to blast open a larger hole into the compound. All nine men knew that speed and surprise were essential if the operation was to be successful.

One man in each team was given the responsibility for laying the explosive charges on the three basement levels below ground and setting the timers. Timing was a difficult decision because they had to give enough time for the man on the lowest basement level to lay his charge and the timer and then make good his escape up three floors. They anticipated that the Iraqi defenders might realise quite quickly that the raid was dedicated to blowing apart their vital military communications system and would therefore move to cut off those men laying the charges.

Four or five minutes would be too long, giving the Iraqis time the stop the timers and save the tower. In the event, they finally decided on just ninety seconds which meant that everything had to go like clockwork for all the men to escape before the charges exploded. That was cutting it fine.

But the SAS raiders had a great surprise. When the men reached the wall they discovered a bomb, presumably from a US warplane, had blasted a hole in the wall big enough for men to gain access easily. A tarpaulin had been placed over the hole. This meant that the SAS men could gain access before having to draw attention to their presence by setting off explosives.

Suddenly, the silence of the night was broken by a single shot. The men looked at each other in anticipation. 'Let's go,' said the Sergeant and the men made for the hole in the wall while those left behind opened fire on the tower. They aimed at any doors and windows visible to them. They wanted the Iraqi technicians and military men inside to believe that there was a large force surrounding the tower and they poured in non-stop metal. The noise was incredible which gave those laying the charges cover to make as much noise as necessary as they made their way into the tower itself and down in to the basements.

But the SAS were in for a nasty surprise. They had not realised that surrounding the tower were bunkers which they now realised were manned around the clock, and suddenly the sound of heavy machine gun fire reverberated around the desert making an awesome noise. That was the worst surprise they could have imagined – heavy machine guns can cause incredible damage and would make escape without death or injury virtually impossible. One bullet from a heavy machine gun will usually kill a man. A way had to be found to silence them if they weren't to take heavy casualties.

The SAS men threw some grenades in the direction of the machine guns and two went silent. It seemed that those manning the Iraqi guns weren't sure of the exact whereabouts of the attackers which did give the SAS an advantage, for Iraqi bullets were firing everywhere with obviously no target in mind. Clearly they were panicking.

Despite the Iraqi fire, however, those men laying the charges had run into the tower and down the stairs to the basement without being spotted. They had even been able to lay their charges and set the timers without being disturbed. Their job was done. Some five minutes after breaking through the hole in the wall those SAS men were spotted at the door of the tower waiting for a lull in the firing before making their escape and sprinting across the open space to the hole in the wall.

Now everyone had to find a way of escaping without being hit. That would be more difficult. The six SAS men guarding the exit had spread out making it appear that there was a large enemy force at their perimeter wall. The metal still honing in on the tower from the SAS men some hundred yards away never ceased which prevented the Iraqis from leaving their bunkers and launching a counter-attack.

Suddenly, the first major charge exploded deep beneath the tower, followed by two more huge explosions which the SAS could feel as the earth around them gently rumbled and shook. They looked up and saw the tower beginning to sway. This was the moment to make good their escape, their one chance to get the hell out of the place alive. And they took it. But there was no rush, no panic.

The SAS men had been well trained. Not for one second did any of them forget the preparation they had undergone for such a real-life eventuality. They knew what each had to do in a fighting retreat under heavy fire and they began their retreat as disciplined as if they were on the barrack square. Some moved back twenty yards while others kept up covering fire. Then they kept up fire while their mates moved and went past them before taking up positions. It went like clockwork – the fire-power coming from their mates back near the Land Rovers ensured that the Iraqi defenders had no chance to raise their heads above the bunkers.

Suddenly from the road came a hail of automatic fire. The SAS had been concentrating so hard on the fire-fight with the tower defenders they hadn't noticed in the dark of the early dawn that some twenty

Iraqi soldiers had taken up a position a hundred yards or so from the Land Rovers and had opened up on the retreating SAS men.

As the SAS turned some of their fire towards the attackers some more Iraqi soldiers had moved towards them firing as they advanced. This was serious. It seemed for a moment as if their escape route would be cut off and the SAS returning from the tower would be faced not only with those firing at them from the flank, but also, eventually, those Iraqi soldiers launching a counter-attack from the tower.

But the SAS men back at the Land Rovers had also noticed the danger and turned their fire on the Iraqis advancing from the road. The SAS soldier firing the M19 grenade launcher turned his attention to the road and fired directly at the enemy which stopped them in their tracks. Indeed, some turned and ran for cover behind their vehicles but more grenades from the M19 slammed into the Iraqi army vehicles, shattering them. The Iraqi fire stopped.

A moment later, the SAS Sergeant screamed 'Go, go, go' to those retreating with him and, as one, the SAS men leapt to their feet and ran like hell towards their waiting Land Rovers. The engines were running and everyone was on board waiting for the nine SAS men to join them. As the men reached the vehicles the Iraqis on the road opened fire again, firing a load of tracer, and the machine guns at the tower also joined in the attack.

As the Land Rovers made their bumpy way across the rough terrain, those on the back kept up their fire on the Iraqis from the road who were now running flat out across the desert in a bid to cut off the SAS escape. But they couldn't run and fire accurately and those Iraqis on the road stopped firing because they would have hit their mates running towards the Land Rovers.

Unbelievably, only one SAS man was wounded – a bullet through the thigh. The injury wasn't life-threatening.

Within minutes both the SAS men and the Iraqis stopped firing as the retreating British Land Rovers put some distance between them and the Iraqi troops. Now they had to keep going and somehow make their way back to base and safety. They realised the Iraqis might well send more troops searching for them and so the only way to escape that possible threat was by keeping away from the roads and driving as fast as possible over the rough, boulder-strewn desert. They intended to take the fastest route possible back to their lines with just a couple of detours to put any following Iraqis off their tracks.

As the Land Rovers bumped their way across the terrain the noise of bombers overhead caught their attention. Then bombs began to drop all around them and they wondered what was going on. They knew they had to be friendly bombers, probably American, but they had no idea why they were dropping their bombs in the middle of an empty desert.

As they came over the brow of the hill the reason became obvious. Immediately ahead of them, a hundred yards or so away, was an extensive Iraqi army camp with hundreds, if not thousands, of soldiers, tents, vehicles, tanks and half-tracks.

The SAS commander leading the short convoy of Land Rovers made an immediate decision – to keep driving and hope that his courage, his effrontery in the face of overwhelming odds, would work. They proceeded at a leisurely pace down the road which ran through the centre of the Iraqi camp. Every SAS man checked his weapons, ensuring all their magazines were full and they held their breath. Unbelievably, there were no guards on duty at either end of the road and no one seemed to take the slightest notice of this convoy of British Land Rovers, carrying no flags and no Iraqi markings, leisurely moving through the camp.

It took three minutes to drive through the camp but the Commander gave no instruction to increase the speed of the vehicles until they were some 800 metres past the last Iraqi troops. Then he ordered the drivers to go like hell. There was just twenty more miles to the forward British position – and safety.

But it was 'Scud busting', of course, that was the most important element of the Special Forces operations throughout the war. Tracing and destroying the Iraqi missiles in the desert was a tailor-made task for Special Forces. The area where the missiles were based was known as the 'Scud Box' and the SAS and the US Green Berets divided this into two areas, the British concentrating on what was called 'Scud Alley' to the south of the road between Baghdad and Amman, the capital of Jordan, and the Americans took care of the north, known as 'Scud Boulevard'.

Roaming around the Iraqi desert in four-wheel-drive vehicles behind enemy lines, both the SAS and the Green Berets began to come across mobile Scuds in their respective areas. At the beginning of the war the Special Force teams would find a Scud on a transporter and would, through communications systems, alert an airbase in Saudi Arabia with the necessary map co-ordinates. A strike plane would be scrambled and directed towards the site. However, the delay between the sighting of the missile on the ground and the arrival of the USAF strike plane would sometimes be from thirty minutes to an hour. Too often the Special Force units on the ground would spot a Scud, transmit the intelligence to base and then wait. By the time the warplane appeared overhead the transporter would have moved on. There had to be another way.

After spotting six mobile Scuds in Scud Alley the SAS men decided that it was a total waste of time risking their skins to track down these missiles and then to see them disappear over the horizon before a USAF A-10 Thunderbolt 11, affectionately called a 'Warthog' or simply 'Hog', arrived on the scene to take them out. They became increasingly annoyed and disillusioned that one mobile Scud after another was parking by the side of a main road for perhaps an hour or two and then moving away before the Warthogs could be scrambled to the designated position. So they decided to take more direct action instead.

From then on, whenever the SAS crews in their Land Rovers came across a mobile Scud, they would prepare a battle plan and go ahead before it moved off. They would usually wait until dark, sometimes attacking just before dawn, when they could safely approach close to the target without being seen.

But the new, improvised battle plans required the SAS teams to adopt a far more dangerous approach. The original plan was for the British and US Special Forces to simply discover the whereabouts of the mobile Scuds and then to call for air strikes to take them out. The new plan, devised on the ground by the SAS units, called for far greater courage and daring in challenging the Iraqi forces guarding the secret weapons with which Saddam Hussein hoped to ignite an all-out Arab–Israeli conflict.

Before reaching the decision to attack mobile Scuds the senior commanders studied the protection the Iraqis were providing for their Scud operators as they moved around the desert. The mobile Scuds were mounted on eight-wheel TEL (transporter-erector-launcher) vehicles and these were accompanied by two Russian ZIL-157V tractors hauling the tarpaulin-covered missiles. Each Scud was accompanied by a petrol tanker and two or three jeeps carrying troops as guards for the Scud and its team of operators.

Small SAS teams of nine men were sent into the desert in groups of three Land Rovers, two with heavy machine guns and a third vehicle with a Milan anti-tank missile launcher fixed in the back of the Land Rover. Their job was to keep out of trouble, search for the mobile Scud transporters and destroy them. The task seemed more like searching for a pub in the Iraqi desert but the teams were in radio contact with base and intelligence was being provided by the pilots of coalition warplanes searching the desert for the mobile missile launchers. Searching for the elusive mobile transporters was, however, made somewhat easier by the fact that the transporters were so heavy they had more or less to keep to the asphalt roads because if they moved off the hard road there was a real possibility of them becoming stuck in the desert sand.

One of the first sightings of an Iraqi transporter and its guards was made by an SAS 'Scud-busting' team after receiving intelligence from a US warplane which had sighted the moving target five hundred miles inside enemy territory. The SAS arrived at the map-reference in the early evening as the sun was setting and discovered the Iraqi troops preparing what appeared to be an evening meal for the guards, missile operators and crew. The SAS took up a position behind a small hillock and watched and waited.

They noticed that no guards had been placed around the Scud transporter and the troops seemed remarkably relaxed, most walking about without carrying any arms and, apparently, no one even patrolling the road. When it seemed the Iraqis had eaten, relaxed and taken to their sleeping bags for an hour or more the SAS prepared for an attack. In darkness, they drove slowly and as silently as possible across the rough desert towards the road and then proceeded at a leisurely pace towards the transporter.

The three SAS vehicles took up their positions some fifty metres from the sleeping Iraqis and then opened fire, filling the air with a crescendo of noise from the heavy machine guns. The Milan fired four missiles, taking out the Scud and the transporter with direct hits while the machine-gunners kept up a bombardment of automatic fire at the Jeeps and anyone one they saw moving. For six long minutes the Iraqis were hit by everything the SAS could throw at them, the machine guns never silent for more than a few seconds at a time. Then, the command 'ceasefire' was yelled and the SAS men surveyed the scene, checking if anyone or anything was moving. The Scud missile and the transporters were a total mess and the Jeeps were smouldering pieces of metal. There was no point in checking out the condition of the Scud and the transporter for they could tell they were now useless. And there was no point in checking if anyone was alive; it didn't matter. It was more important to get out of the place and leave the mess to some shocked Iraqi force to discover. It was time to go.

As well as attacking the mobile Scuds, the SAS units would also take the same aggressive action against any communication or radar installations they came across as they moved remarkably freely across the desert area of Scud Alley. The great majority of these attacks proved highly effective, enraging the Iraqi high command.

In Scud Boulevard, where the US Special Forces were active on the ground, it was a similar story of courage and professionalism. Delta Force squads made up of six-man teams operated independently and were given specific areas of Scud Boulevard to cover in their search for the mobile Scuds.

The US Special Forces were nearly always parachuted into the designated areas, dropped by MC-130 Combat Talon aircraft, which are fitted with infrared imagers enabling aircrews to visually identify targets and checkpoints at night. These imagers could pick out any Iraqi deployments on the ground at potential DZs, or dropping zones, which meant that no US Special Forces would be ambushed as they parachuted into the desert. They were also fitted with terrain avoidance radar and an inertial navigation system providing accurate navigation to unmarked DZs.

The Delta teams were taking no chances. Their black balaclavas were replaced by desert flop hats and they painted their faces yellow and brown rather than black. They wore pale, 'no sweat' bandanas and camouflage scarves and chocolate-chip-patterned camouflage jackets and trousers. Most Delta soldiers carried a Colt M16A2 assault rifle or a Heckler & Koch MP5 sub-machine gun. They also carried Beretta handguns fitted with the latest silencers, which are so efficient that only the sound of the hammer striking the firing pin can be heard when a shot is fired. In case Iraq resorted to chemical or biological warfare, each soldier carried a gas mask strapped to his left leg and a complete NBC suit in his rucksack. He was also equipped with decontamination gear, extra water, food and medical supplies. In addition, every member of the team had a pair of AN/PVS-7 night-vision goggles, which turn the desert darkness into a greenish but very visible day-like scene. By amplifying the available light they enable man-sized targets to be recognised at a distance of one hundred yards without moonlight. When there is some moonlight the range is increased to about one hundred and fifty yards.

The communications expert in each six-man team had fitted a burst-transmission capability to his radio so that with a touch of a finger a pre-coded message could be flashed instantly back to base. This allowed hostile direction finders no time to get a fix on the team's position.

The United States Special Forces used more sophisticated gadgetry in their quest to find and destroy Saddam's mobile Scuds. As the Delta teams in their Jeeps criss-crossed the desert they would stop frequently to check their hand-held satellite systems, taking intelligence from the small screen's luminous digital readout which arrived as coded signals from the Global Positioning System in orbit above the Iraqi desert. As new intelligence arrived the Delta teams would move position, sometimes by just a mile or so, sometimes fifty or a hundred miles away.

Like the SAS, the Delta squads would stay some distance away from the main roads, remaining in positions hidden from sight by the Iraqi vehicles travelling back and forth. But they would seek outcrops from where they could see through their binoculars – Steiner 7 x 40s – whether any of the convoys travelling along the asphalt roads included Russian missile transporters with their precious cargo of Scuds.

On one occasion, after days and nights of fruitless tracking of hundreds of Iraqi convoys driving up and down the highways, one Delta team spotted a slow-moving transporter some half mile distant. Ten miles further on, the Scud transporter and its small accompanying Jeeps came to a halt and the operators began preparing the Scud for launching.

This intelligence and the exact location was immediately passed on by the troop's Satellite Communications System through central communications to the United States Air Force command at the Royal Saudi Air Force headquarters near Riyadh. Some twenty minutes later five US A-10 Thunderbolts – or Warthogs – arrived above the Scud transporter and circled some distance from the Iraqis.

Those guarding the Scud also went into action, turning their guns to face the approaching threat, getting a fix on the Warthogs before opening fire. The Warthogs approached at twelve thousand feet and levelled off eight thousand feet above the road, firstly firing their missiles and then dropping their bombs as they came closer to the Scud. Once they had dropped the bombs they immediately swooped back up into the sky to escape the Iraqi air-defence missiles.

For ten minutes this life-and-death game was repeated by the USAF Warthogs, diving, levelling off, firing and then climbing to escape being shot down by an anti-aircraft heat-seeking missile. For their part, the Iraqis seemed incapable of organising the defence, running all over the place desperately trying, usually without success, to blast off their missiles. Their anti-aircraft gunners did carry out heroic work sitting at their posts while the Warthogs' missiles and bombs rained down on them, but, they were having no luck because the aircraft never came within their range of fire. Instead, the convoy took a pasting and by the end of the onslaught the Scud, the transporter and all the Iraqi vehicles were simply wrecks. And there was no sign of life. Once again, the Delta team saw nothing to gain by going to inspect the wreckage or check for any survivors so they quietly disappeared again in the desert to search for another Scud.

This Delta team had a number of other opportunities to locate mobile Scuds, and each time they were able to bring up the USAF Warthogs to the scene and repeat their Scud-bashing exercise. Sometimes the USAF sent in Tornado aircraft to carry out these deadly attacks. And nearly every time a mobile Scud was located, the attack that followed ended in the destruction of the missile, the transporter and most of the remaining vehicles.

Sometimes, however, things did go wrong.

On one occasion a Delta team were making their way on foot by moonlight in search of the origin of some noisy engines. The team had been awoken during the night and identified the sound of both jet engines and other diesel engines which couldn't have been that far distant. They went to investigate.

But as they walked silently in file the lead Delta man suddenly shouted 'Shit', and seconds later they heard the sound of automatic gunfire from the ridge to which they were heading. Unable to see the enemy, the Delta squad hurled some half-dozen grenades over the ridge in the hope that this Iraqi unit was just a few men at the far reaches of the area they were guarding.

The Iraqis replied by hurling half-a-dozen grenades back over the ridge, right into the midst of the Delta squad. Two exploded, one injuring one of the squad, two more were picked up by the Delta boys and hurled back, and two more sailed harmlessly overhead, exploding some distance away, causing no harm. While the wounded man was taken back some twenty yards and placed behind a rock, the other Delta boys threw some more grenades over the ridge. They were seriously worried that if the Iraqis came over that ridge firing machine guns from the hip then there would be one hell of a close combat struggle.

The lead man reported that he believed there to be some twenty to thirty Iraqi guards against the five fit soldiers and their one injured comrade.

The Delta boys were ready, their magazines full and their weapons cocked; they were waiting for action. They didn't have to wait long. Suddenly, on top of the ridge some twenty yards from their position, six Iraqis stood for a split second before opening fire on the US troops. But the US troops were in the better position, difficult for the Iraqis to pinpoint their positions in the dark. The Iraqis, on top of the ridge, were an easy target, silhouetted against the night sky. Three fell almost immediately. The rest turned and jumped back from the ridge.

Instinctively, three of the Delta team knew this was the moment to strike hard and fast. Their task was to seize the initiative and try to drive the Iraqis backwards down the hill. The three men raced to the top of the ridge and threw themselves down before opening fire on the twenty or so Iraqis who were looking up at them. The Iraqis also opened fire but their bullets were screaming over the heads of the prone Delta boys while the Iraqis were taking casualties from the raking machine gun fire of the American MP5s.

The rest turned and fled down the hill as the Americans continued firing until it was obvious the fight had gone out of the enemy. Now, they had to make sure they could get their wounded soldier – hit in the shoulder – back to their Jeeps and away to a safe area where a USAF chopper could cas-evac the wounded man. They made it without any further interference, but it had been a close run thing.

As a result of such raiding parties by the Coalition Special Forces, the Iraqis changed tactics. In an effort to keep the raiding parties at bay, mobile Scuds were provided with APCs and sometimes light tanks for greater protection. It was believed that these armoured vehicles, equipped with heavy machine guns and other weapons, would be able to take on and destroy the Special Forces in their small, light four-wheel-drive vehicles.

The Iraqis also sent a score of strong mobile units in APCs and light tanks, usually led by jeeps with mounted medium machine guns, into the Scud Box to hunt for the Coalition Special Forces. These Iraqi units would patrol the main roads and tarmac tracks along which the Scud transporters had to drive. But they were too late, for the damage had been done. It is believed that some ten Scud transporters were wrecked by the Special Forces and, occasionally by the Warthogs, thus all but putting an end to the menace of the mobile Scud. The great majority of the fixed Scud emplacements had been targeted by satellite photography and taken out by the Warthogs.

Within seven days of the start of the Coalition's air war, the last of the Scuds had been fired on Israel. The destruction of the Iraqi missiles had been a fast, efficient and brilliantly executed operation by the combined Special Forces of Britain and the United States. It was also a magnificent opportunity for Special Forces to show their true value on the world stage to the planners, military strategists and defence experts from around the globe who were watching this very modern war with the utmost interest.

AFRICAN INFERNO

'LUCY, LUCY, LUCY.' The call sign for the assault sounded calm and clear through the troop commander's headphones as he sat in the noisy chopper hovering above the roof tops of the hot, dusty African city, waiting for the order to go into battle. Wearing flak jackets and desert fatigues, Staff Sergeant Matt Eversmann and his team were crammed into the Black Hawk with their weapons, each man carrying fifty pounds of ammunition and assault gear.

Below them men, women and children were running around, waving and pointing to the squadron of helicopters – Little Birds and Black Hawks – swooping in from the north. The billowing smoke and acrid smell of tyres burning in the streets were everywhere. That sight worried Eversmann because he knew the tyre-burning was the signal to summon the gunmen when trouble flared. And those tyres had been burning for some time.

He scanned the streets below, wondering what sort of reception committee would be waiting when his men hit the ground and went into action. But as he looked up he took comfort, for circling above were the P3 Orion spy plane, three OH-58 observation choppers and the communication satellites. This was a state-of-the-art United States military operation, equipped with the latest sophisticated weapons, communications systems and highly trained combat troops.

Following Eversmann's strike-force team were seven elongated troop-carrying Black Hawks: two carrying elite Delta Force assault teams and their ground command, four bringing in US Rangers and one carrying a search-and-rescue team. On the ground waiting in support were nine wide-bodied armoured jeeps – Humvees – with roof-top machine guns, filled with more Special Forces soldiers; and three five-ton US Army trucks to bring out the assault force and any prisoners. A total of one hundred and sixty troops manned this armada of planes, helicopters and land vehicles.

On the ground waiting for the US assault were the enemy – a band of Somali militiamen armed with Kalashnikov AK47 automatic rifles and the odd machine gun, dressed in nothing but colourful shirts and cotton trousers, and many in bare feet. Also gathered in the streets – as if attending a carnival – were women in colourful dresses and scores of wide-eyed children, all eager to see what was about to happen.

The crack US Special Forces had been sent to Somalia to sort out the African warlords who had reduced the nation and its capital, Mogadishu, to a lawless state in which rival gangs of thugs ruled by the gun, looting, shooting and killing one another and making life a living hell for the impoverished population.

This mission was designed to attack a gathering of top Habr Gidr clan leaders, who controlled most of the armed gangs and who were in turn led by the infamous warlord Mohammed Farah Aidid. The task was to capture two of Aidid's senior lieutenants and any other clan leaders and imprison them on an island off the Somali coast.

US Intelligence had established that the clan leaders were to meet that day – Sunday, October 3 1993 – in a house in the heart of Aidid's territory, the so-called Black Sea neighbourhood, the most closely guarded place in Mogadishu. The target building was only two hundred yards from the busy Bakara Market and across the road from the five-storey Olympic Hotel. L-shaped and on three floors, it was given privacy by a high stone wall that surrounded both the building and its courtyard.

The first two Little Birds – McDonnell Douglas AH6 Apache helicopters with four men on board – landed in an alley on one side of the house and the teams leapt out, throwing thunderflashes to scare but not injure anyone who was nearby. They raced through the iron gates, across the courtyard and up a small flight of steps into the house, shouting at everyone to lie down. But the place seemed empty. They burst into a shop attached to the building, smashed their way into a warehouse behind the shop and found that empty too. They had the wrong target.

Both teams ran back into the street just as the Black Hawk troop carriers hovered low overhead and the Delta Force and Ranger commandos fast-roped down, dropping through the cloud of dust and sand that filled the air.

As the choppers took to the air again, urgent radio messages were going back and forth from the ground commanders to those in the skies. Undoubtedly the teams had landed at the wrong place. They were some one hundred yards from the target. Worse still, the Somali militiamen had begun hitting back, firing at random into the dust cloud, making life unpleasant for the commandos. And Eversmann knew that this was the third such fuck-up in as many weeks; the intelligence on the ground had been shit. Every time US Intelligence personnel had been given information about which building was being used by Aidid's men, it had turned out to be wrong.

The reception this time was fiercer than ever. Within minutes thousands of Somali women and children had poured into the street, drawn by the clattering helicopters and the sound of gunfire. The women began erecting makeshift barricades around Bakara Market as the men opened fire with all the weapons in their possession. Those in the command helicopters above the town watched helplessly as jeeps and open trucks raced towards the market, overflowing with gunmen rushing to the defence of their town.

This riposte wasn't new to the Special Forces. The Americans had been in Somalia six weeks and each and every week they had undertaken similar missions in the capital, sweeping in low in their choppers, beating up the town, scaring the people with low-level flights. Sometimes they landed, bagged some Habr Gidr personnel and then headed out of town – sometimes by chopper, other times by truck. But they had not yet landed any big fish.

While Eversmann's force held off the gunmen, the three other Delta 'bricks' – four-man groups – found the correct target building and burst in. They raced through the large house, hurling deafening thunderflashes and shouting for everyone to lie on the floor. On the first floor they found two dozen Somalis, among them Omar Salad and Mohammed Hassan Awale, two clan leaders, and, more importantly, Abdi Yusef Herse, an Aidid lieutenant. But as they were putting plastic cuffs on the prisoners, rapid machine-gun fire sprayed the wall and ceiling of the room. Experience told the Delta men that the weapon was a US M60 and that those shooting at them were their own comrades.

One team took the prisoners and moved out of the room on to a flat roof at the front of the building. Suddenly they were confronted by shots from AK47s. They returned fire and then another US M60 opened up, pouring automatic fire on them. Sergeant Paul Howe, the man in charge of the brick, radioed the Delta Force ground commander. 'One of our teams are firing their M60 at us,' he said calmly. 'Will you tell them to stop before someone gets killed.'

What amazed the US Special Forces throughout their operations in Mogadishu was the bravery of the Somali women and children, who saw it as a duty to act as human shields for their gunmen. They would stand in front of a militiaman as he fired an AK47 because they were convinced the US forces would not open fire on women and children. On one occasion a US Ranger and his mate saw a Somali gunman lying on the ground firing his AK47 at US troops. The man had the barrel of his rifle between the heads of two women as they lay on the ground in front of him. Another time a Somali gunman made three teenage girls slowly walk in front of him towards the US soldiers as he fired shots. There was no way the US forces would shoot at a gunman in such a situation, so on both occasions they threw thunderflashes at the group. Shocked and scared, the women and the gunmen simply ran away.

Despite the occasional burst of fire from the choppers overhead, the Somalis kept up sporadic firing at the Delta and Ranger bricks as they searched for Aidid's men, moving in and out of buildings and watching their backs for fear of gun-toting Somalis. But already the Somali irregulars were massing behind their woman and children in the north of the city and moving towards the US forces.

Relentlessly, the Somali militia edged ever closer to the US positions, firing behind their women and then darting for cover so that the Delta and Ranger troops could not get a crack at them. Things were looking desperate for the US troops because the Somalis – now numbering several hundred – were within fifty yards and their firepower was increasing. Shawn Nelson, the M60 machine-gunner of one of the teams, had to make an agonising decision. The Somalis were now coming towards him and his mates, running, stopping, firing and then repeating the process as they moved ever closer. He waited until they were only twenty yards away and then opened fire at them. At the same time a Little Bird, armed with rockets, came swooping low down the road behind the attacking Somalis and let fly two rockets into the crowd. A sheet of flame hit the mass of people, killing and injuring dozens and scattering those who survived.

Satisfied that the mission had been successful and that his Special Forces had captured three key personnel, General William Garrison, the Texan in command of the operation that day, checked the intelligence coming from the satellites and aircraft in the sky and ordered the Special Forces in the town to wrap up the mission and return to their airfield base in the Humvees and five-ton trucks.

But as the Special Forces, with their prisoners, began to drive away from the market and the target house, they were not only confronted by thousands of people standing across the road and screaming abuse, but were being fired on from both sides of the street. They could also see the trails of RPGs – rocket-propelled grenades – homing in on their vehicles, and that was truly worrying. A direct hit on either a Humvee or a truck would smash straight through the side and take out a number of men.

Exposed and unable to speed out of the city, a number of Ranger and Delta soldiers were hit by gunfire. And then an RPG exploded into one of the five-tonners, bringing it to a sudden stop. Three wounded Rangers tumbled out, their legs shattered.

But Sergeant Eversmann and his men were still holed up, desperately trying to 'cas-evac' – casualty evacuate – a man on a rescue helicopter. But the wounded soldier, unable to help himself, had fallen out of the chopper and had to be roped back in. All this took time and during those vital minutes the number of Somali gunmen had suddenly increased. Small-arms fire and RPGs were coming in all around Eversmann, his team, the medics and the rescue helicopter.

Somehow they had to keep the gunmen – now in their hundreds – far enough away to make their escape. With their M16 rifles, rocket grenades and an M60, the small band kept up withering fire, while the Somalis kept their distance, not willing to risk their lives in an all-out attack.

As the rescue chopper took off with the wounded soldier, Eversmann and his men looked up to see a Black Hawk attack chopper coming to their aid, its guns blazing and rockets firing at the attackers as it skimmed the roof tops. The gunmen ran for cover and Eversmann and his men relaxed, confident now that they would be able to make their escape.

Then they saw the Black Hawk suddenly disappear from view. It had taken a direct hit and crashed to the ground.

Many Rangers saw Black Hawk *Super-Six-One* go down, watching in disbelief as it took a direct hit on the tail from an RPG. With the noise of a thunderclap, the tail rotor exploded; the engine began to make an odd noise; the chopper slowed, stopped and began to spin towards earth, twisting faster and faster. As *Super-Six-One* clipped the top of a house and flipped over, the rotor blades were ripped off. It came to a crunching jolt of a stop in an alley below, throwing a cloud of dust into the air.

News of the crash crackled over the US military radio network as everyone reported to Command what they had seen. No one knew how many on board the chopper had been killed or injured, but the very fact that a Black Hawk had been downed over the city by a single RPG had shocked the troops. Even the tough, cool Special Forces were shaken by the incident. Suddenly no one was safe and the mission had taken on a totally different shape. The US Rangers and Delta Force soldiers were now in a real battle, the like of which none had ever experienced. Their lives were under threat and they would have to work hard and fast, relying on their training and professionalism, to get out of the place alive.

A teenage Somali youth named Aden told what happened after the Black Hawk crashed: In *Black Hawk Down* Mark Bowden wrote:

I saw two US soldiers stagger out of the wrecked helicopter, one of them carrying an M16. I was frightened, so I ran away and hid under a VW

parked nearby. The soldier with the M16 came and stood next to the car. Across the road was a Somali gunman also with an M16. They seemed to see each other at the same moment and both fired their guns. The Somali's rifle jammed and the American ran over to him and shot him in the head.

The American turned away and a big Somali woman in a bright dress came running down the alley charging towards him like a goat. Suddenly the American turned, saw the woman running towards him and fired. She fell to the ground dead.

Then I saw more Somalis come running and shouting and shooting at the American soldier. He dropped to one knee and began to fire back, hitting one Somali with every shot. They were falling down in front of him. Then the soldier was hit but he still carried on firing.

I heard the noise of another helicopter and looked up to see a small one, coming in low. It landed on the ground not far from me. Soldiers jumped out. Some began firing at a crowd of Somalis nearby while two others ran towards the crashed helicopter.

The noise was deafening. Above the clatter of the helicopter blades I could hear grenades and firearms being fired almost non-stop. The bullets were coming from alleys, houses and buildings all around me and the American soldiers were firing back. I saw soldiers carrying two wounded men back to the little chopper and then it took off. I just lay on the ground under the car not daring to move. I was scared I was going to die.

To General Garrison and his aides back at the Joint Operations Center (JOC) watching the debacle unfold on CCTV, the odds on getting all of the one hundred and sixty men out safely were narrowing by the minute. He ordered another Black Hawk, with a fifteen-man team of Combat, Search and Rescue soldiers (CSAR) on board, to the immediate vicinity of the downed chopper. But he knew that the CSAR team, simultaneously tending wounded men and fighting a rearguard action against a howling mob of gunmen screaming for revenge, would have a tough time.

As he planned his next move General Garrison watched the CCTV screen, which showed countless Somalis running towards the crash site from every direction. Nearly all were armed. Some were racing in from further afield, hanging off cars and light trucks, desperately keen to join the battle. Plumes of black smoke from burning tyres hung over the entire area. Garrison was confident that his crack Special Forces could hold out against several hundred ragged gunmen, but he feared that Aidid's own squads of trained combatants would be a different matter. They had good equipment and proper training. And if they arrived at the crash site before the CSAR team could get in and out again, Garrison feared his troops would be in an impossible position – pinned down, short of ammunition and surrounded by armed men.

There was only one answer. If he was to save the lives of those dedicated soldiers he needed numbers. A squad or two was no good against hundreds of armed men who wanted revenge. He knew full well how the Somalis had mutilated the bodies of the US Special Forces who had crash-landed in another Black Hawk only weeks before.

Word was flashed to the commanders of the 10th Mountain Division, part of the 75th Ranger Regiment, to mobilise immediately and get to the crash site by whatever means and in the shortest possible time. Garrison had decided to do all in his power to save his men, but he knew this would be the greatest test of US Special Forces.

Rangers and Delta soldiers on the ground near the market were ordered to make their way in their Humvees and trucks to the Black Hawk crash site, some half a mile distant. But Eversmann and his men, including a number of wounded, were pinned down and had no vehicles. They were ordered to make their way to the crash site on foot, but with their wounded comrades that was impossible. Some couldn't walk and Eversmann had only five who could still fire a weapon. It seemed as though they would all go down fighting, overwhelmed by the massive numbers of the enemy.

All at once an eight-vehicle convoy of Humvees and trucks came round the corner, also heading for the crash site. On board were Rangers and Delta men. 'Shit, it's good to see you,' Eversmann said with a sigh of relief. Manning the heavy-duty machine gun on the turret of the first Humvee was his mate, Sergeant Mike Pringle.

As Eversmann and his men crossed the road to the vehicles, Pringle kept up a stream of fire with his .50-calibre gun, making sure no Somali gunmen would take a potshot at them. They piled the wounded on top of those already lying in the back of the vehicles and the convoy took off towards the crash site, while the Somali gunmen again took to the streets and peppered the convoy with their AK47s and the occasional RPG.

It was while they were travelling through the hostile streets that their radio gave them another piece of devastating news. Control told them: 'We just had another Black Hawk go down to RPG fire south of the Olympic Hotel. We need to get everyone off the first crash site and get down to the second crash site and secure it.'

The men who heard the news exchanged nervous glances, knowing that this information changed everything. Escaping a barrage of fire had proved difficult, but this new order meant they would be lucky to escape with their lives.

The convoy was now moving through incessant fire from all sides and the casualties were mounting. The vehicles were taking direct hits, and it was RPGs that were causing the damage. Although the great majority missed their target, when one of the rocket grenades did hit home the effect was devastating. The men inside the vehicles were simply blown apart. The convoy was still a long way from safety and the wounded were moaning and crying out in real pain.

Though the Humvees and trucks were overloaded with assault troops, prisoners and the wounded, they had to make two stops before getting out of Mogadishu. Those in command of the convoy wondered how they would be able to pick up more wounded and assault troops from the first crash site and then make their way once more through withering gunfire to the second crash site to rescue more of their comrades. Only then, with all the vehicles crowded to capacity, could they make a dash for safety. It was a tall, if not impossible, order.

The next fifteen minutes would be hell for those in the convoy. They were all fully aware that there was no chance any of them would survive if they were captured by Aidid's ragtag army. It was 'kill or be killed'.

It was then that the convoy got lost.

In the heat of the moment everyone made mistakes. It began when the lead vehicle, containing the commander of the convoy, Colonel McKnight, was hit and shrapnel bit into his arm and his neck. It was obvious that those back at base had little or no idea what the convoy was going through. They continued to issue orders to go to the first crash site and then to the second crash site as though it was some bus trip.

Colonel McKnight seemed disorientated and lost in the maze of streets, and asked Command for instructions and guidance. But apparently his position was misunderstood and, as a result, he was given the wrong directions.

Worse would follow. A second convoy was dispatched to the second crash site in a bid to rescue those trapped in and around the Black Hawk and as a result further incorrect directions were given to the 'lost' convoy, which was now driving around the area being shot at from both sides of the street and facing heavy gunfire at every crossroads.

As Mark Bowden wrote in his no-holds-barred classic *Black Hawk Down*: 'So the convoy now made a U-turn. They had just driven through a vicious ambush in front of the target house and were now turning around to drive right back through it. Men in the vehicles behind could not understand. It was insane! They seemed to be *trying* to get killed.'

But the chaos on the ground continued as crew men in several helicopters overhead tried to direct McKnight and his convoy to the correct spot. Once again incorrect directions were given. The incoming gunfire grew more intense and the Somali gunmen sensed that the convoy was lost and that they were within an ace of a famous victory – hijacking a US convoy and killing more than a hundred Special Forces soldiers.

In one RPG hit, three men were catapulted out of the rear of a Humvee on to the road as the grenade exploded inside it, igniting the petrol tank. All three suffered horrendous injuries. Those inside, who did not suffer the full force of the blast, received shrapnel wounds, broken bones, torn ligaments and serious gashes. The convoy stopped to pick up the wounded men and, with no thought of their own safety, three soldiers leapt from the Humvee, dashed back down the road in a bid to save their wounded mates. The shooting rose in a crescendo as they dragged the shattered bodies of their comrades back to the vehicle.

The machine-gunners manning the turrets in the Humvees all took direct hits and were too badly wounded to continue. Some were removed from their cockpits unconscious; others at death's door. But, as soon as one gunner was knocked out, another man would take his place, knowing that he was putting his life at risk. But none of them gave such fears a second thought. It was their duty and they would carry it out no matter what the consequences. Their heavy-calibre machine guns were doing a fantastic job keeping the Somalis at bay, and so these weapons became the prime target of the militiamen.

Those men in the convoy still capable of fighting now faced a new and even more serious problem; and one that could not be easily overcome. In their efforts to hold back the Somali gunmen they had had to expend a huge amount of ammunition, and by now it was running dangerously low. As well as having to take the magazines from their wounded pals lying on the floor of the vehicles, they had to cut down on the number of rounds they fired. While they realised this would make life easier for their pursuers, there was no alternative. They began shooting only at the Somalis within close range, making sure that every shot hit its mark, taking out another gunman.

But even that ploy had no real effect. As soon as one Somali went down, someone else would run forward, pick up the fallen weapon, take aim and renew the firing. There were now thousands of Somalis, screaming and yelling, running down the streets and firing their weapons at the convoy as it snaked hither and thither, desperately trying to locate the second downed Black Hawk. With an extraordinarily naive disregard for their own safety, the Somali gunmen now lined both sides of the streets and fired as the convoy drove past. As a result, many found themselves shooting at their own side across the street, and some of the rounds that missed the US vehicles hit their fellow Somalis, killing quite a few.

For the convoy the situation had become bleak. But the bravery of those US troops shone through the disaster that was unfolding all around them. They relied on their training, their professionalism, their pride in the unit and their camaraderie.

As those in the convoy finally caught sight of the second crash site, the driver of one of the trucks, Private Richard Kowalewski, who was called 'Alphabet' because no one could pronounce his name, took a bullet in the shoulder. But he refused to let anyone else drive the vehicle because it was his responsibility. Private Clay Othic, who had been shot in the arm himself, was trying to apply a pressure dressing to his friend Alphabet's shoulder when suddenly the vehicle was hit by an RPG. The grenade roared into the cab, severing Alphabet's left arm and entering his chest. Unbelievably, it didn't explode on impact but imbedded itself in Alphabet's body, the fins sticking out of his left side under his missing arm. The nose was sticking out of his right side. He was unconscious but still alive.

As a result, the truck crashed into the vehicle ahead, which, in turn, smashed into a brick wall. All the other vehicles behind them came to a halt. Othic was knocked unconscious by the impact and woke to find someone screaming at him, 'Get out, get out, the truck's on fire.'

Dazed, Othic fell out of the truck and then remembered young Alphabet unconscious but still at the wheel with the grenade right through him. He stumbled back to the blazing cab and managed to drag his mate out. Medics at the scene put Alphabet in the back of one of the remaining Humvees, but they didn't believe he would survive his horrific ordeal.

The convoy set off again, but now it was travelling more slowly and making an easy target for the enemy gunmen. The tyres on most of the vehicles had been blown out and they were running on flats. The vehicles were designed to keep going on flat tyres, but at half the normal speed. In addition, a number of engines had taken direct hits. The convoy would have difficulty in making it back to safety, let alone finding the second Black Hawk and taking on a whole load more wounded and assault troops.

Soon the vehicles had passed out of the market area, leaving behind many of the gunmen, the vast majority of whom had been on foot. But there was still occasional incoming fire.

The convoy had now been meandering around Mogadishu for forty-five minutes while the Somalis sprayed them with live rounds and hit them with RPGs, scoring many direct hits. Colonel McKnight decided it was useless to try to reach the crash site. Of the seventy-five men in his convoy, nearly half had been shot or injured by shrapnel, and eight were dead or dying. He radioed Control, demanding that the convoy be permitted to head back to base, but they refused, telling him it was vital he return to the crash site at once. But he and his soldiers were in no condition to go on a search-and-rescue mission in the heart of enemy territory, crawling along in their smashed-up vehicles, their ammunition all but gone and half of the men incapable of holding a gun.

McKnight told Control that it was vital the casualties received medical attention a.s.a.p., and the convoy started off for Headquarters. But they weren't safe yet. As they moved off, Chief Signalman John Gay, of the SEAL commandos, took the lead in a Humvee. Riddled with bullet holes and with the engine smoking, the vehicle was crawling along on the rims of three wheels. There were eight wounded Rangers and a body in the back. On the bonnet was another wounded Ranger.

Then, down the road, they saw a group of Somalis running around having set alight a barricade made of two underground petrol tanks stacked with old furniture and rubbish. Gay knew that if he tried to drive off the road and round the obstacle, there was every likelihood the Humvee would come to a grinding halt. But he put his foot down on the accelerator and crashed through the burning barricade, throwing everyone to one side of the vehicle as it nearly turned over. Now they were heading for safety – or so they thought.

But the gallant men in the beaten-up convoy had been forced to leave behind some seventy Rangers and Delta comrades. These troops were now in a desperate situation, fighting for their lives against a howling mob of thousands of armed Somalis bent on killing the Special Forces soldiers holed up in pockets of resistance between the original target house and the first downed Black Hawk.

Those in command back at base could see on TV screens and hear on the radio the desperate plight of their men and they had no means of rescuing them before the Somalis moved in for the kill. The crash site was in danger of being overrun. If that happened, the commanders knew, every US soldier at the scene would be slaughtered. The Somalis would take no prisoners.

At the same time the men trapped in the city were convinced that rescue was on the way. They had no idea of the true situation, nor did they have a clue as to the number of dead or of wounded comrades who had somehow fought their way back to base.

The one hundred and fifty men of the 10th Mountain Division had been thrown into the fray in a desperate attempt to rescue the doomed soldiers. They raced out of the HQ in their Humvees and five-ton trucks, hell-bent on reaching them before the Somalis overran their positions. But as soon as the 10th reached the outskirts of Mogadishu, they faced the same impenetrable problems as the other convoys. Their progress was blocked by barricades hastily erected in the patchwork of streets and then, when they were forced to slow to a crawl, they were subjected to torrents of fire from automatic weapons.

Chief Warrant Officer Mike Durant in Black Hawk *Super-Six-Four* and his co-pilot, Chief Warrant Officer Ray Frank, had heard the report of the downing of *Super-Six-One* on their radio from another chopper pilot who witnessed the crash. To them, to Staff Sergeant Bill Cleveland and Sergeant Tom Field, and to everyone else in the helicopter, the downing of a Black Hawk by an unruly mob armed with AK47s and a few RPGs seemed incredible. But it gave them a warning of the dangers ahead.

Ordered to maintain a low, sweeping circle at roof height over the battle area where the Black Hawk crashed, the men in *Super Six-Four* could see everything going on below – the other circling US choppers, the columns of Rangers moving along the dusty streets, the crashed helicopter and the bands of armed Somalis converging on the stricken aircraft and its crew.

Having made a few sweeps around the chopper, Durant felt a thump and knew they had been hit. Quickly he checked the instrument panel; everything seemed to be OK. But, flying behind him, was another Black Hawk and the pilot reported to Durant that oil was pouring out of his tail rotor. Reluctantly, Durant headed back towards the airfield. Suddenly the tail rotor exploded and the airframe began to vibrate. A second or so later the Black Hawk went into a spin and, without the tail rotor, Durant was unable to stop it. As the ground rushed up towards the men in the eight-ton chopper, Durant and Frank pulled every lever they could think of in an effort to pull it out of its spin and level off. Somehow, when it was only a few feet off the ground, it levelled out and landed almost flat.

The crash landing was very, very lucky. The Black Hawk had come down in a shanty-town area where there were no brick buildings, and its super-tough shock absorbers had managed to take the impact of the crash without breaking up. Everyone on board had taken a hell of a battering, but they were alive and there were no broken bones.

Unfortunately, however, the TV screens at HQ did not show the exact location of the downed helicopter and so those in command could not give the CSAR team an accurate map reference. Worse still, the TV shots from the helicopters flying high in the sky showed bands of armed Somalis now running through the town towards the second crash aircraft.

In response, General Garrison ordered a Quick Reaction Force (QRF) to be assembled. In the lead Humvee would be Rangers and Delta troops who had just returned from Mogadishu after rescuing a soldier who had fallen out of a helicopter seventy feet to the ground. But they needed more men, and so dozens of support personnel, including cooks and communication specialists, volunteered for the rescue mission.

As they sped out of the HQ in a trail of dust, another helicopter pilot flying over Durant's crashed Black Hawk reported that the Somali gangs were within a hundred yards of the chopper and closing in fast. Trying to hold back the swarming Somalia militiamen converging on the site were two Little Birds and another Black Hawk, flying low over the scene, strafing the angry crowds, throwing down grenades and firing bursts of automatic fire at them.

Chief Warrant Officer Mike Goffena, piloting the Black Hawk, aimed for the Somalis holding RPGs because he knew they could do the most damage, not only to the crashed chopper but also to the three helicopters trying to hold off the gunmen. So he flared his Black Hawk very low over the crowds, creating clouds of sand and dust and driving the gunmen back. But the angry Somalis firing the RPGs wouldn't pull back. Instead they held their ground and tried to fire a grenade at the chopper just feet above their heads.

In turn, this exposed the RPG gunners, and the snipers on board the Black Hawk took them out with well-aimed shots. But as soon as one Somali had been felled, another would run out from the crowd, pick up the RPG and take aim. Small-arms fire was splattering the Black Hawk every time it flew low over the throng and Goffena feared one lucky shot might prove disastrous.

Sergeant Jeff Struecker would lead the volunteer QRF convoy heading for the second crashed Black Hawk, with two Humvees in front, followed by three five-tonners, and two more Humvees bringing up the rear. On top of the forward Humvees were heavy-duty .50-calibre machine guns and the three trucks were filled with Rangers with M16 rifles, M249 portable machine guns and loads of ammunition. Any Somalis who opened up on this convoy would be met by a hail of lead. But the convoy hadn't driven a hundred yards out of the airfield when, to their amazement, they were met by a hail of gunfire.

What worried Struecker was the fact that every time the convoy came across Somali small-arms fire there were always one or two RPGs as well. He knew that if any vehicle in the convoy received a direct hit from an RPG there would likely be heavy casualties.

Within minutes of leaving the compound, Struecker realised that the convoy would never get through to the stricken Black Hawk, so heavy was the enemy fire. He knew that the crashed Black Hawk was only a mile or so away from his position, but the problem was getting there through all the incoming shit. After talking to Command, Struecker went for the only possible option – to circle the city and approach Durant's chopper from the other side.

Meanwhile, above the confusion and chaos on the ground, Mike Goffena and his crew could see Colonel McKnight's battered main convoy travelling at a snail's pace back to base, carrying the dead and wounded and those Special Forces soldiers still firing away at the Somali gunmen. They could also see the Somali gangs closing in on the second downed Black Hawk and the QRF led by Struecker trying to battle their way through to the stricken helicopter.

On board were Master Sergeant Gary Gordon and Sergeant First Class Randy Shughart, two highly trained Delta snipers with combat experience. Goffena had identified a field some thirty yards from the crash site and the intervening area was dotted with shacks. He believed that Gordon and Shughart would be able to keep the mob at bay while those Special Forces in the downed chopper would be able to make their way the short distance to Goffena's position, where they could be picked up by helicopters.

The senior officers in the air and back at base knew the two Delta snipers would be taking an extraordinary risk, pitting their firepower against hundreds of armed gunmen. But neither

Gordon nor Shughart could leave their mates to die at the hands of angry Somali gangs without first trying everything to rescue them. Finally, Command relented – they could have a go.

As planned, Goffena put down the two men in the field, dropping them when just five feet off the ground. The two Delta snipers had no idea of the condition of the survivors at the Black Hawk crash site and decided to make an immediate dash to the helicopter before any Somalis even knew of their arrival on the ground. They figured that if they were there with those guys they would together be able hold back the Somalis creeping ever nearer the crash site.

When they saw the state of both the chopper and the crew they were glad that had made the decision to help – whatever the consequences. Mike Durant and Ray Frank had been knocked unconscious by the force of the crash, but both men happened to come round just before Gordon and Shughart arrived on the scene. Durant knew he had a broken leg and guessed his back was also broken. The pain was intense. Frank had also injured his back and his left tibia was broken, but he had dragged himself away from the chopper to get a better line of fire and to draw the incoming fire away from his mate. Both men had suffered crushed vertebrae in the crash.

For some time after they were downed, Durant had remained conscious, and later he recalled firing off shots with his pistol every time a Somali came too close. In that way he had managed to keep them away, but he knew there would soon be many more baying for their blood. Then, as the minutes dragged on, he had lost consciousness.

Frank had spent the time away from the chopper but in a good position, with his back to a high wall, which meant that he could see the gangs of Somalis but they couldn't get behind him. This allowed him to keep firing whenever they crept too close, and they would back off. But he knew that couldn't last for long.

During roof-top sorties over the crash site Goffena could see that Gordon and Shughart had sorted things out on the ground, arranging the crew of the Black Hawk in a perimeter around the downed chopper. They all seemed to be in secure positions with some cover. And all waiting desperately for the arrival of the volunteer QRF.

The pilot was playing an heroic defensive game, hovering over the Black Hawk while his co-pilot picked off any Somali holding an RPG. But they were taking the most enormous risk. One grenade on target from an RPG would bring them down instantly, probably killing them. They seemed prepared to take that risk if it meant they were helping to give their mates on the ground a chance to live.

By now the streets and alleys around the crash site were filled with Somalis converging on the crash site. Goffena kept trying to scatter them, flying his Black Hawk just feet off the ground, but the Somalis would lie down and then jump to their feet after the chopper had passed overhead. He could see that the troops on the ground could not last long without help and there was still no sign whatsoever that the QRF was at hand. The situation was becoming truly desperate.

Goffena knew that he was risking his life and the lives of his crew by flying at ground level in an effort to keep the howling mob away from his mates around the crashed Black Hawk. But he also knew that if he didn't hold the Somalis off, then those soldiers would be dead meat. So, with no thought for his own safety, he kept up his remarkable one-man battle, repeatedly hurling the Black Hawk down towards the crowd, skimming them in a cloud of sand and dust and then pulling out, turning in the air and diving down to ground level once more, making it difficult for the Somalis to move forward to the crashed chopper. And it worked.

Then, like a thunderbolt, an RPG struck.

The grenade struck the right side of the Hawk, killing the engines stone dead. The rotors ground to a halt and the helicopter's alarms screamed out. Goffena looked around him. His co-pilot, Captain Yacone, was slumped in his seat unconscious; Sergeant Brad Hallings, a Delta sniper, had one leg crushed; but Sergeant Paul Shannon and Sergeant Mason Hall, sitting in the back, were uninjured.

Goffena knew they were going down and his head was clear enough to remember that he had practised for this a hundred times on a flight simulator. He knew that he must keep the bird level, get the nose up before impact with the ground and stay cool. Somehow he managed all three. He selected a narrow alley in which to bring down the stricken chopper and, when only feet off the ground, pulled back sharply on the stick.

But, as Goffena and his crew braced themselves for the impact of a crash landing, the chopper suddenly responded. Instead of landing on the ground, it somehow pulled out of the slow dive and flew on. Goffena wasn't sure what engines and rotors were somehow still working, but he found that by keeping the nose up he could fly on at roof height. Within seconds they were out of the city. Ahead, Goffena saw the new port facility, which was guarded by US Marines, and hoped the damaged Black Hawk would keep going long enough to reach safety. He skimmed the perimeter fence and brought down the chopper. As soon as the wheels touched the ground it keeled over and came to a crunching halt. They were safe. As the pilot clambered out of his seat to check the condition of his mates, he was thinking of the possible fate of Durant and his crew, whom he felt he had failed.

But Durant and his men were still holding out. After Gordon and Shughart had been among the men, checking their injuries and fighting capability, Gordon took up a position in which he could confront the main thrust of the Somali militia. Hunkered down and with enough ammunition to hand, he was quietly confident that he could deter the gunmen until help arrived.

Shughart moved the injured Durant some fifteen feet from the fallen chopper so that he could sit with his back to a tree and get a better view of any approaching Somalis. Then he rummaged inside the bird and came out with two M16s and plenty of ammunition. He knew that he would need the help of every man capable of firing a weapon if they were to halt the Somalis' advance until the QRF arrived.

At that moment Durant heard an anguished cry from Gordon's position. Then there was silence. Shughart went to investigate and came back with his mate's Car-15, a carbine based on the M16 infantry rifle but with a shorter barrel and collapsible butt. He loaded the weapon and handed it Durant, along with a fresh supply of ammunition and wished him luck. He said nothing of Gordon's fate but Durant knew he must be dead.

Shughart called up on the survival radio and then positioned himself on the other side of the chopper from where most of the fire was coming. Within minutes, however, there was a tremendous onslaught of incoming fire and Durant heard Shughart cry out. Silence followed. Now he knew he was totally alone and unable to move, with only a Car-15, an M16 and his own pistol to keep the Somalis at bay. He had no illusions. This was it.

Durant fired at a face which was only yards away; then another ten or more approached and he stopped firing. It was useless. He wondered what they might do. They didn't shoot him, but tore off his clothes, kicked and punched him and then dragged him away blindfolded. Then he fainted.

Back at the first crash site, where *Super-Six-One* had come down, Chief Warrant Officer Cliff Wolcott, the pilot nicknamed 'Elvis' because of his cool personality, was still waiting for the ground convoy to arrive and rescue them. Around Wolcott's downed Black Hawk were some thirty US Special Forces, both Rangers and Delta boys. They were having little trouble keeping the groups of enemy gunmen at bay. Some of them were worried as they saw bands of Somalis edging towards their positions, but they had calm, experienced leaders. They did feared for their safety, however, when they heard over the radio that the relief convoy had run into trouble and no exact ETA – estimated time of arrival – could be given. As dusk was falling, orders were given to move the wounded into a house adjoining the crash site. Those capable of firing a gun took up defensive positions and waited for the convoy and the cover of night.

In command was Captain Scott Miller, the Delta ground commander, and various small groups of Special Forces soldiers held four other positions down the road from the crash site, where thirty men were gathered. Some were wounded, some were medics, others were Black Hawk crew and the rest were Rangers and Delta men who had arrived on the scene shortly after the chopper had crashed.

But the Somalis were getting more daring and creeping closer to the crashed Black Hawk. The five groups of men were now totally encircled and gunfire was coming in from every direction. One by one the defenders were taking hits from small-arms fire and the medics were having to work hard taking care of all the injured. Occasionally someone was hit by an RPG – a sight that brought a shudder to all who witnessed it.

As darkness fell the order was given for everyone to move back towards the Black Hawk and take up defensive positions in the house next to it, where all the wounded could be treated in one place. In that way Captain Miller hoped that he could save the lives of more of his men and make life easier when the rescue team finally arrived.

When the men had all gathered in the new base by the chopper a quick check was made on ammunition, water and IV (intravenous) bottles. All were dangerously low; all would need a re-supply drop, and fast.

General Garrison discussed the situation with his senior officers. All agreed that the need to re-supply was vital. All agreed too that there was every chance the rescue helicopter would be brought down by enemy fire. But he still gave the order to go.

Super-Six-Six's pilots, chief warrant officers Stan Wood and Gary Fuller, watched as their Black Hawk was loaded with kitbags containing ammunition, water and IV bottles. The idea was to fly directly to the crash site and hover as low as possible over the site while the kitbags were pushed out of the aircraft. Within minutes the chopper was over the target, and as it descended out of the night sky, machine guns, AK47s and RPGs opened up on it. The men inside the base house were surprised how close the enemy were to their position.

The Black Hawk hovered for about thirty agonising seconds while the kitbags were thrown out into the night. It was getting riddled by bullets; the rotor blades were taking direct shots and so was the main engine and gearbox. But somehow no RPG found its target and, having dropped its precious cargo, *Super-Six-Six* lifted off and accelerated away, making it safely back to the airfield. The damage was serious, but, more importantly, those men surrounded by a ferocious enemy had been re-supplied and they could now resume the fight to save their lives and the lives of their wounded comrades. They all knew it would be a long, exhausting and anxious night for everyone.

The Little Birds kept making darting gun runs over the Somalis, swooping low, firing off in bursts of automatic fire and then accelerating away from the small-arms fire and the even more dangerous RPGs.

Those officers high above the battle area could see that there was an urgent necessity to get all the Special Forces men into one consolidated position where they could defend a designated perimeter for as long as necessary. The commanders watched the battle below through infrared and heat-sensitive cameras that sketched the area in black and white. They were all aware that there was every chance that the one hundred and thirty or so soldiers on the ground, including the wounded, might have to defend their positions throughout the night before a large enough force could be assembled and sent in to rescue them. A major offensive, including tanks and armoured personnel carriers (APC), was planned.

But those on the ground were not happy with the orders being radioed over to them by commanders in the air and back at base. They all understood that it would be better for all the men to be consolidated in one defensive position near the crashed Black Hawk, but the commanders didn't accept that plan was highly dangerous, as two groups of men would be risking their lives as they moved there. To some of them the plan seemed near suicidal.

Nevertheless, Captain Miller decided to lead the way and ordered four Delta soldiers to spring out of their position, across the main road and into the house adjoining the crashed Hawk. As they sprinted out of their courtyard a wall of rapid gunfire greeted them. All four came rushing back to the spot they had left seconds before, diving back into the courtyard. They were all convinced they would never have made it the forty feet or so across the road.

As the night wore on the shooting died down, but the Special Forces men could take no chances. Everyone knew they had to stay awake and alert, ready for any attack from any direction. Occasionally the silence would be broken by the distant hum of a Little Bird on its way to make another gun run down a street. Minutes later the shattering noise of automatic fire and rockets would startle everyone as the little chopper with just two crew flew fast down the street, firing at the enemy positions. Always the Somali AK47s and machine guns would open up, and usually an RPG or two would be fired. But luck seemed to be with the Little Birds.

The rest of the time, the quiet of the night was unnerving for the trapped soldiers. Most were convinced that the Somalis were sending small teams to probe their positions, finding out exact numbers. Others believed the Somalis might send suicide squads armed with RPGs who were prepared to take ridiculous risks in order to attack and kill or wound their enemy. A hit by a well-aimed RPG could be catastrophic in the confined spaces they were defending. And they already had fifteen men with serious injuries, some of them life-threatening. The injured could not easily be moved.

Time seemed to stand still as the men strained to listen for any movement of encroaching Somali gunmen. Occasionally, if one was seen edging too close, a burst of fire would quickly make him change his mind and retreat. Everyone knew that the Rangers and Delta boys coming to their rescue would have one hell of a job holding off a concerted attack over several hours.

Shortly after midnight the soldiers on watch heard the first faint rumblings of tracked vehicles seemingly only a few hundred yards away. They had been informed by radio that a relief force was on its way, but they were also told that the Somalis seemed determined to make that journey as difficult as possible.

As the convoy edged closer the men could hear the thunder of its guns and the noise of a heavy machine gun, which sent the fear of God into the Somali gunmen. The men holed up had been informed that the rescue convoy of tanks, APCs and Humvees numbered nearly one hundred and that the order had been given to take no prisoners. For the first time in many hours there were smiles on their lips. They were going to make it.

Some time after five in the morning, as the orange sun was rising over the roof tops and the last of the wounded were placed in the APCs, the order was given to return to base. After the 10th

Mountain Division soldiers had boarded the APCs, the Malaysian drivers simply sped off, unbelievably leaving behind those Rangers and Delta boys they had come to rescue. These were the men who had just spent the last fifteen hours fighting a desperate rearguard action against overwhelming odds.

They shouted and yelled but no one in the APCs could hear above the roar of their engines and the rumble of their half-tracks. There was only one thing for the Special Forces men to do – run for their lives. They had to cover just over a mile, carrying their packs, their weapons and spare ammunition, and the Somali gunmen were waiting for them. They were moving targets. As they ran they fired at anything that moved. They fired at windows and doors, and down alleyways, anywhere they thought a gunman might be hiding. Incoming small arms-fire was increasing by the minute and a number of men were hit.

These guys hadn't eaten for some twenty hours, surviving on a few sips of water. In the sustained firefight they had lost many of their comrades. They were shattered, physically and mentally, and now, incredibly, they had to run the gauntlet once again, and face an avalanche of fire like the one that had felled and injured so many of their mates.

News of what followed on that hellish day in Mogadishu reverberated around the world as Somali television showed dead US Special Forces soldiers, stripped half naked, being dragged around the city behind jeeps and trucks while the local people jeered, cheered and revelled in their hour of glory. For President Bill Clinton, his entire administration, the Pentagon and every American watching, those news pictures were the most humiliating and degrading images ever seen on US television.

Everyone was asking the same brutal question. How could the US Special Forces, backed up by the world's most powerful army, navy and air force, be routed and humiliated by a few thousand Somali gunmen?

Eighteen Americans had been killed and dozens badly wounded. And the pride of Delta Force, the Rangers and the SEALs had taken a battering.

The whole operation, scheduled to last an hour, should have gone like clockwork. Well armed, well trained, well disciplined and with excellent helicopter support, the Special Forces should have been able to capture a couple of military aides of a Somali warlord with ease. They didn't, and those eighteen American soldiers paid the ultimate price.

Yet this extraordinary episode had proved one thing – that America's Special Forces had the stomach for a fight even when battling against great odds. It showed wonderful camaraderie among the men on the ground, who risked their lives to save their colleagues. But for their courage and ability to stay calm and disciplined in a desperate situation, the casualties might have been far worse.

CHAPTER 11

THE SOVIETS' GRAVEYARD

THE DEATHBED OF invading armies for two hundred years, Afghanistan has an awesome reputation for any nation that dares to set foot in this rugged, mountainous country where warlords and tribal gunmen have nearly always held power.

Once a distant part of the Persian Empire, Afghanistan was occupied by many different peoples during its early history, but the rule of law never extended much further than major towns and cities. During the nineteenth century Russia took over much of the country and persuaded its countrymen to take control, but the local warlords had no wish to be ruled by anyone, particularly Russians.

Some warlords turned to Britain for assistance because the Russians seemed ready to move south into British-held India, which then included Pakistan. In 1939 the British sent in the Redcoats to bring Afghanistan into the British Empire, but the Afghanis had different ideas. Rather than live under the discipline of the British Crown, they rebelled and, in 1842, wiped out the entire British garrison in the capital Kabul. The British withdrew, but returned in 1878, taking the two main cities, Kabul and Kandahar, and subduing the Afghan warlords. Britain only wanted to keep a token force in Afghanistan as a buffer state to dispel any Russian idea of driving south into India and beyond. However, World War One effectively ended the British presence in Afghanistan and the country lapsed once again into a backwater in which local warlords changed alliances, fought battles and lived off the opium trade.

Decades later, in the 1970s, the Soviet Union's Secret Service, the KGB, decided for various important reasons that Afghanistan should become a part of the Soviet Union in all but name. The rationale behind this policy was that Afghanistan had become a back door for trade for many of the southern Soviet satellite countries, which had become disenchanted with the Soviet system of government. The KGB also saw the poppy fields of Afghanistan as a lucrative source of illegal funds for its own department, which was becoming increasingly starved of income by the close-fisted Soviet treasury.

In this way the KGB would be in a strong financial position to keep its power base within the Soviet system intact and functioning efficiently without any strings attached. The KGB leaders were ecstatic when the Soviet politburo agreed to their plan for Afghanistan. During the previous decade they had come to believe that the KGB's power and influence were declining in the face of a stronger, more powerful United States, which it had been incapable of halting or even slowing. Now they could see a highly lucrative, more powerful future for the department, with their finances guaranteed.

They began by organising the removal from power and, in 1973, the assassination, of the King of Afghanistan, Mohammed Zahir Shah. Slowly but surely the Soviet Union eased troops and civil servants into the country and came to dominate the main cities. Five years later the new Afghan republic, led by the Soviet puppet President Amin, signed a peace treaty with the Soviet Union.

But the Soviet forces were unable to dominate any of the

autocratic Muslim warlords who held power outside the main cities and towns and, as the Soviet Union increased its grip on the country the warlords became resentful and eventually troublesome. One of the main reasons for the breakdown was the KGB's ambition to wrest control of the poppy market from the Afghan warlords, who were equally determined this would not happen. The opium trade provided the funds which gave the warlords their power. In short, this money paid for guns and ammunition for their local fighters.

The more the Soviet forces moved out into the country to subdue the warlords, the more resistance they met. The warlords called on their fighters, whom they had armed with the latest Soviet AK47 rifles, and the local Muslim peasants, who made up some eighty per cent of the rural population, to resist the pagan Russian invaders. Slowly the conscript Soviet army found itself the target of increasing guerrilla attacks and realised it had no stomach for the fight.

The KGB leaders came to the conclusion that they would have to rule the country through another puppet leader, and that would mean removing the warlords and their power bases and subduing the Afghan farmers and peasants by force of arms. But first it would be necessary to remove President Amin and his close advisers quickly and clinically. The Spetsnaz, the Soviet Special Force, was given the task. The KGB's new ideas for the virtual annexation of Afghanistan, including the assassination of Amin, were rubber-stamped by the Politburo.

Already the KGB had the run of Darulman Palace, the heavily fortified home of President Amin, so that the logistics and planning presented no foreseeable problems. The initial raid on the Afghan forces defending the palace's grounds and defences was given to a crack Soviet airborne battalion.

The defenders were taken completely by surprise because they believed the Soviet paratroopers falling out of the sky above

Kabul were friendly forces on an operation. They were unprepared for the ferocious onslaught that began as soon as the leading company of paratroopers landed in the palace grounds. The Soviet paras were taking no prisoners and the Afghan forces who stood their ground and fought back were wiped out within a couple of hours.

Only when the ground was secure did the Spetsnaz, under the command of Colonel Boyarinov, Commander of the KGB sabotage school at Balashika, move in for the attack on the palace itself. The three hundred-strong Spetsnaz assault force, which included officers of GRU, Soviet Military Intelligence, met strong resistance.

Boyarinov gave orders to the Spetsnaz that every single person in the palace, including President Amin, had to be liquidated. His instructions were that no one should be permitted to survive to tell the world what really happened that day. The KGB's plan was that the world would be told that the President had died as a result of a palace revolution and that the Soviet Special Forces had arrived too late to save the lives of the President and his advisers.

But Amin's personal bodyguard of trusted loyal fighters proved no pushover. They realised what was happening as soon as the Soviet paratroopers began landing in the palace grounds, for they knew there was no planned exercise. When shooting began a few minutes later they were in no doubt that this was a planned attack on the President.

But the two hours it took for the Soviet paratroopers to secure the grounds and kill or drive off the defenders gave those inside the palace time to throw up barricades, arm everyone in the palace capable of firing a rifle, stack ammunition in rooms surrounding the President's office and prepare for the attack.

The crack Spetsnaz force directed flame-throwers at the main doors in a bid to scare the defenders into throwing down their arms. One or two non-combatants did come out with their arms raised, only to be cut down by automatic fire as soon as they were sighted by the Spetsnaz forces. This initial shoot-to-kill order was noted by the defenders inside the palace, who now realised that the battle would be a fight to the death, and so they fought with incredible defiance and courage. The Spetsnaz forces knew the layout of the whole palace, but they had no idea there would be some fifty well-armed, highly disciplined tribal fighters prepared to lose their lives in defence of the President.

As the Spetsnaz forces smashed through the fire-damaged main door they were met by a torrent of automatic fire which drove them back and left some ten of their number dead on the palace floor. They changed tactics, racing around the building, throwing grenades through windows in an effort to confuse the defenders as to where the next assault would come.

During the next frontal assault some Spetsnaz soldiers succeeded in gaining access to not only the main hall but also one of the rooms. This enabled them to bring in more troops, so that the defenders were now having to fight on two fronts. Even so, the Spetsnaz found themselves pinned down, unable to press further into the palace. Every time they tried to gain access to another room they were met with a fusillade of automatic fire, making progress impossible.

When the Spetsnaz brought up rocket grenades in addition to the flame-throwers, the defenders withdrew to the first-floor rooms where President Amin and his advisers had been placed for their own protection. The President and his close circle were now totally surrounded, with all phone links cut by the attackers.

But the President's loyal fighters were determined not to surrender and to kill as many Soviets as possible. As more Spetsnaz forces gained entry to the ground floor the defenders on the landing above kept up a barrage of automatic fire, making it impossible for the attackers to climb the stairs. Every onslaught by the Spetsnaz forces was beaten back and their casualties began to mount. The rocket grenades and flamethrowers, the second a favourite among Soviet forces for clearing houses, seemed not to be having the usual effect.

Indeed the tribal fighters, armed with their new AK47s, began to pin down the Soviet troops inside the downstairs rooms. Their wide field of fire made it impossible for the Spetsnaz troops to break out of the rooms without taking serious casualties. And then the Spetsnaz attack seemed to falter and the blaze of gunfire became more like a trickle. The Spetsnaz were running out of ammunition.

It seemed that the Soviet planners had seriously miscalculated the number of loyalist defenders and the amount of ammunition President Amin had stored in the palace for just such an eventuality. The miscalculation was to cost the lives of many Spetsnaz troops, who by now could only try to hold their positions and prevent the tribal fighters from launching an attack on their precarious positions on the ground floor.

The Soviet airborne forces were still holding the palace grounds, but they had handed over all their spare ammunition to the Spetsnaz troops. It seemed that a stalemate had developed inside the palace and so Boyarinov decided the only way to break down the President's defences was to bring in more Soviet troops and artillery to surround the palace and destroy the entire building and its occupants with heavy artillery fire.

Boyarinov took an opportunity during a lull in the fighting to race out of the main door of the palace. He had not reached the bottom of the steps before he was shot dead by machine-gun fire from his own forces. He had forgotten the strict orders he had given personally to the airborne troops – to kill anyone they saw leaving the palace building.

More ammunition, and many more rocket grenades and hand

grenades, did arrive for the Spetsnaz forces and, with assistance from the paratroopers, they began to inflict serious losses on the defenders. The tribal fighters withdrew to the rooms immediately surrounding President Amin and his advisers as they sustained more casualties, and eventually they too began to run out of ammunition. The end was fast approaching but some half-dozen fighters made one last attempt to turn the tide. Armed with sub-machine guns and full magazines, they raced, screaming, down the central stairs, blasting away at the Soviets in the reception area below. Several Spetsnaz troops were killed and wounded in the ferocious attack but it was to no avail. The handful of brave tribal fighters were all killed.

It was only then that the Spetsnaz forces began in earnest to move from room to room, searching for any defenders. Before entering each room they would throw in a couple of grenades, wait for any sound and then rush in, sub-machine guns firing. There was never any intention of arresting or detaining anyone found on the premises. Every single person they came across, man, woman or child, was gunned down in an appalling orgy of killing. In this way the Spetsnaz went through the entire palace until not a single person was left alive.

When the Soviet troops entered the room where President Amin and his advisers were sheltering there was apparently no discussion and no questions were asked. The President and the seven men with him were simply gunned down, their bodies riddled with scores of bullets. Now the Soviet Union had almost total control of Afghanistan. But it would not last.

Soviet troops began to target the warlords and their tribesmen with increasing ferocity as the KGB tried in vain to wrest control of the opium trade from the thousands of peasants who grew the poppies in both small patches and large areas of land wherever the seeds would flourish. But the peasant farmers, backed by the local tribesmen and their warlords, had no intention of letting their harvest be stolen by the Soviet forces.

The KGB's plan was to subdue great swathes of the rugged countryside with force of arms and a continued military presence. There was no attempt to win the hearts and minds of the local population and, as a result, the Soviet troops quickly became a hated and despised army of occupation. And then the warlords, supported by America's Central Intelligence Agency, which saw a chance to humiliate the Soviet Union, went on the offensive.

It was during the following seven years that the CIA encouraged, funded and gave arms and ammunition to all those Afghanis prepared to take up arms against the Soviet troops. The CIA threw its support behind the Muslim clerics of the Taliban and their political leaders, who seemed to be growing in power and influence. Indeed the Taliban leaders had stepped into the political vacuum caused by the overthrow of President Amin and his cabinet and slowly they became the centre of anti-Soviet activity. At one time during the middle 1980s, secret CIA flights were bringing in arms and ammunition to Taliban guerrillas three times a week. And, much to the CIA's delight, the Taliban were knocking hell out of the dispirited Soviet forces.

Throughout the 1980s the Soviet forces, mostly conscripts with no appetite for war in such a cold climate, found themselves facing a continuous guerrilla war being waged by tribesmen with far greater knowledge of the inhospitable terrain and a fierce determination to drive them out of the country. A total of fifteen thousand Soviet troops were sent home in body bags throughout an eight-year war in which the Soviet forces had been seen as brutal, cowardly and ineffective. Shortly after Mikhail Gorbachev became President of the Russian Republic, following the collapse of communism, he began talks to extricate the Soviet forces from Afghanistan. He signed a peace accord and by 1989 all Soviet forces had been withdrawn, leaving the devoutly Muslim Taliban authorities in complete control of most of the country.

There was rejoicing throughout Afghanistan, but not for long. The strict mullahs were intent on turning the country into a model Muslim nation, living under strict Sharia law, which calls for the flogging of a woman for committing adultery or even revealing an ankle, and the amputation of a man's hand for theft, a leg for robbery with violence. These laws were not received well by the Afghani tribesmen, who had always lived a more free and easy life. The Taliban also decreed that the growing of poppies was prohibited by Sharia law. Penalties for growing the crop and harvesting the opium included flogging, amputation and death. At a stroke the Taliban had taken away the peasant tribes' main source of revenue.

In an epilogue to this salutary story of the disastrous Soviet campaign in Afghanistan, the United States forces arrived in the country towards the end of 2002, intent on rooting out and killing those responsible for the September 11 attack on the World Trade Center, in which some three thousand innocent people were killed. The United States sent its Special Forces in the vanguard, backed by an alliance of Special Forces from Britain, France, Germany, Australia, New Zealand and the Netherlands. In fulfilment of the political element of the allies' plans, the Taliban were removed and, overseen by the West, Afghanistan is presently emerging as a fledgling democracy. But it is too early to evaluate the success of the massive counter-terrorist mission initiated by the United States, which, by some estimates, could last for years. There is, however, no doubt that the role played by the Special Forces will be crucial.

CHAPTER 12

SPECIAL FORCES TRAINING

A N IMPORTANT FACT about Special Forces personnel
throughout the world is that, without exception, they are
all volunteers. This immediately marks them out as remarkable
and daring young men who have a deep love of adventure but
who also display the determination, courage and discipline which
are the making of a first-class soldier.

Once a young man has taken that all-important decision to
volunteer, the hard part follows almost immediately. From the
moment he arrives at the Special Forces camp for training he
quickly learns that this will entail tough discipline, shattering
physical exercises and nerve-racking problems. But the
volunteer will also discover an amazing *esprit de corps*, the
remarkable extent of his own capabilities and a rightful pride in
his own achievements.

He will not have realised that he was accepted for training by
reason of the fact that the officers and NCOs who interviewed
him saw that he was a man of maturity, common sense, and
ambition but also someone with intelligence and a sense of
humour. All five qualities are essential if the volunteer is to make
the grade as a Special Forces soldier. Without just one of them he
would not be able to play a full role in an elite fighting force.

Those in charge of recruiting are also looking for a

characteristic which is not as immediately visible. When volunteers are being put through their paces before selection the recruiting personnel – all experienced Special Forces soldiers – are looking at a volunteer's 'attitude'. The volunteer must have the character to accept the toughest training schedules and to learn things about himself from the experience. It means finding reserves of energy and determination when he is on the point of exhaustion; never complaining, whatever the appalling circumstances he finds himself; and having the ability to smile and laugh at himself at the end of an horrendous exercise.

The training of most Special Forces volunteers the world over is much the same. All will have been given some three months' basic training with other military units. During this the recruit gets fit, learns to drill and obey orders, as well as receiving basic weapons training in grenades, rifles, sub and light machine guns and sometimes mortars and anti-tank weapons. He also learns basic field skills.

And then he volunteers to join Special Forces and life takes on a very much tougher hue.

Some of the world's Special Forces units organise pre-recruit training, where the volunteers can sample the life they will lead if they complete their training. This gives those who will be training them a chance to weed out any young men who they believe not to be the right type of person for their specialist unit or who might not fit in with the spirit of the training school. It is during this pre-recruit training that a volunteer's sense of humour can be judged. Those who have a distinct deficit in this area are often failed at this stage of recruitment but are not usually given the reasons. Most volunteers who don't make the grade seem to know instinctively that they will fail, and sometimes they know the reasons too.

The selection of volunteers for the famed 22nd Regiment, Special Air Service, is a prolonged business which tests the would-be recruits in many different ways. Volunteers for 22 SAS are split into groups of twenty and they begin with ten days' fitness training, hard runs and lots of press-ups. They are then bussed to the Brecon Beacons, in Wales, for ten days of map reading and long, hard cross-country marches which end in a muscle-wrenching forty-mile endurance march that must be completed in twenty hours. Each volunteer carries a fifty-five-pound Bergen rucksack and a rifle. Those who don't make the finishing line in twenty hours are RTU'd – returned to their unit – immediately.

Two methods of discovering the faint-hearted recruit are often employed by the senior NCOs, both of which some recruits see as below-the-belt tactics. One is the offer of food from an army Land Rover towards the end of a gruelling hard march. If the hungry recruit even looks like accepting food he is immediately RTU'd. The other ploy is to tell recruits at the end of a long march with full pack that there is still another five miles to go. Anyone who as much as groans at the news meets the same fate. Of course, there are always some who reach the decision that the SAS is not for them, and they may leave at any stage.

After this introduction the volunteers are launched into a fourteen-week intensive training course, but throughout this time they can still be sent back to their unit at any time if those in command believe they won't make the grade. Those who pass this phase then take the British Army's parachute course. Success here leads to combat survival training, which includes surviving interrogation and methods of escape. Only after passing this stage is the volunteer a fully-fledged member of the SAS and entitled to wear the famous sand-coloured beret and the SAS cap badge with its winged dagger and motto 'Who Dares Wins'.

But that is not the end of the affair.

The new member must now learn more skills, including field medical care, various languages, Morse code, pistol shooting and the use of explosives. At the same time he will become skilled in climbing, long-range overland navigation in Land Rovers, free-fall parachuting and boat work. Only after he has passed in all these disciplines will he be capable of taking his part in any SAS operations. The training will have taken two long, hard years of application, discipline and physical endurance. But the look of pride on the faces of all those who pass the test to join 22 SAS shows why the men put themselves through such a living hell.

Despite the vigorous selection procedure and the care taken to select only those who the recruiters believe will make the grade, the failure rate is still remarkably high. Most Special Forces training units demand the same basic high pass mark because they understand what will be demanded of the volunteer when he enters service with the unit and, as a result, between seventy and eighty per cent fail at some stage during the training process.

There is, of course, no punishment for failing the training and being returned to one's unit. However, those three letters, RTU, represent a most unwelcome outcome for any Special Forces volunteer. For they instil in him a feeling of deep humiliation, a realisation that he is not one of the select few whom he was dying to join, an understanding that he is a reject among those he admired and wished to emulate, and failure is often followed by a deep depression that can last weeks or months. Indeed some failed volunteers have gone on record as saying that being RTU'd from Special Forces training was the worst moment in their lives, an experience they have never been able to forget. Some go so far as to admit that their basic ambition to achieve something in life in which they could take pride has been stripped from their psyche.

Once accepted into the SAS the recruit will concentrate on one of four specialist troops – the Boat Troop, Mountain Troop, Mobility Troop or Free Fall Troop. This means that the SAS can meet the demands of Special Forces operations in any terrain in the world. The specialisation also means that each troop contains individual specialists, radio men, first-aid and medical orderlies, snipers, signallers and linguists, the last of whom who may be proficient in English, German, French, Italian, Spanish, Russian, Arabic, Mandarin Chinese and Malay.

During the past thirty years, the SAS has also placed an ever-increasing emphasis on anti-terrorist tactics. Today one squadron is always kept on standby in Britain for Counter-Revolutionary Warfare. At any one moment this is on call for immediate duty anywhere in the world. Indeed standby officers routinely listen to the news broadcasts each and every hour so that they can identify at the earliest moment a possible opportunity for the SAS. Quite often, as for example when the news broke that the Argentines had landed on the island of South Georgia, in the Falklands, in 1982, the senior SAS officer on duty with the standby squadron immediately phoned the Ministry of Defence, offering their services. They don't like to wait to be asked but prefer to volunteer whenever there is the possibility of action.

Nearly all recruits begin their training with a feeling of confidence and excitement, even though they know the going will get tough. Most recruits even start off by admiring or liking the senior NCOs who will be their taskmasters for the next few months. But it doesn't usually take long for those officers to be seen as pitiless, harsh, inflexible, ruthless bastards as, with never a smile crossing their lips, they put the recruits through their paces.

It is only later that the recruits come to realise that these senior NCOs are in fact their true friends, respect them as first-rate soldiers and admire them for their professionalism and integrity. At the end of most training schedules the recruits believe that the officers who have trained them are the greatest people they will ever know in the Special Forces. And they will thank them, time and again, for pushing them hard and teaching them how to react when things get really tough – in battle conditions.

Undeniably, these senior NCOs are the making of Special Forces recruits the world over. All have experienced Special Forces operations at the limit and they know and understand how tough life can be when under fire in conditions which favour the enemy and disadvantage their own unit. It is these experienced men who are the backbone of the elite forces and the better trained and more experienced they are the better Special Forces units will eventually operate in battle.

All those who undergo the training schedule in a Special Forces unit hear, along with the myths and legends, some true horror stories about mishaps suffered by trainees. These would turn any mother's hair grey if her son had volunteered to join one of the elite forces.

The United States Marine Corps has earned itself a reputation for being tough on recruits. The examples are endless. One recruit was locked in his locker and lighter fuel poured through the vents and then set alight. Another was kept for three hours in freezing water. Yet another had a bayonet pushed through his biceps. One was made to run around the square with full pack until he collapsed from exhaustion. Another had his head repeatedly pushed into a bucket of ice water until he collapsed. Then there was one who was stripped naked and made to stay outside his billet in freezing conditions. Another poor soul had to continue press-ups until he collapsed. Yet there have been no reports of any volunteers dying while undergoing such punishments. Nevertheless, all recruits who witness or simply hear of such horrific events do realise how tough the training regime can be.

The US Ranger Department of the Infantry School at Fort Benning, Georgia, established in 1951, trained officers and NCOs initially in hard physical exercise but also in patrolling, ambushes, raids, airborne operations and leadership skills. Jungle and mountaineering training was also included. But the emphasis was changed to individual training. And a target was set: 'To produce a hardened, competent, small-unit leader who is confident he can lead his unit into combat and overcome all obstacles to accomplish his mission by requiring them to perform effectively as small-unit leaders in tactically realistic environments.'

As one senior Ranger NCO put it: 'When you've earned your Ranger tab [a curved yellow-on-black strip bearing the single word 'Ranger', worn on the left shoulder] you'll know you'll be able to do anything, even if it's something you've never done before and know absolutely nothing about. That doesn't matter. You're a Ranger, you can do anything.'

The official history of the Rangers Brigade states:

To produce a realistic environment, the stress of combat is simulated by hunger, lack of sleep, constant pressure, and all in a gruelling physical setting. The long training day usually lasts from 0500 hours to 0200 hours, and because it is similar to battle conditions makes judgement difficult. The short rations – one or two small, processed meals a day – add further problems. The average weight loss per student is thirty pounds... At the end of the course the student is in the worst physical shape of his life.

The Ranger training programme differs in some points from that of the SAS. Ranger recruits are assigned buddies, the idea being that if one falls behind the other will help him. No Ranger does anything alone, for he and his buddy work as a team. The buddies stay together until they graduate, but often Ranger buddies remain friends for life, regardless of rank.

At the end of the training period a new phase was introduced in the 1990s, entitled Ranger Stakes. This provides a test of students' abilities in eleven separate tasks in the areas of light infantry weapons and communications. The first three tests involve the M60 machine gun and include loading, range-finding and maintenance. Task number four is to set up an M181A1 Claymore mine and detonate it. Tasks five and six involve communications, sending radio messages and coding and decoding. Tasks seven, eight and nine involve everything a recruit must know about the US Army's basic weapon, the famous M16 rifle, including maintenance, correcting malfunctions and cleaning. Task ten involves hand grenades and the final task is the maintenance and firing of an M203 grenade-launcher.

Perhaps the greatest difference between the SAS and the Ranger recruiting programmes is that when an SAS recruit is failed and RTU'd this is the end of his time with 22 SAS, whereas in the Ranger programme if any recruit fails a phase of training he can decide to re-enter the next training programme and have another try. The Rangers also employ specific and intense desert, mountain, jungle and swamp phases of training at various Ranger compounds – Fort Bliss in Texas, Dahlonga in Georgia and Elgin Air Force Base in Florida.

At the completion of training less than thirty per cent of those who began the training course pass out as full members of the US Rangers and only fifty per cent of the intake will have managed to get halfway through the course.

The SAS are looked up to by many armies across the world as the epitome of Special Forces units, occupying a position which all should strive to emulate. However, the SAS believes that the toughest Special Forces training is undertaken by the US Navy SEALs. In contrast to Air Force and Army Special Forces in the Rangers Regiment and Special Forces groups, SEALs are 'generalists', meaning that although each will have a speciality in intelligence, submarine operations, weapons, engineering or communications, as soon as a SEAL unit goes off to war every man has to be capable of doing the job of anyone else in the unit.

To earn SEAL flippers takes twenty-six weeks of exhaustive training which is described as 'the most challenging and brutal learning experiences anybody can ever have'. And it's true.

The official title – Basic Underwater Demolition/SEAL

programme, or BUDS – conceals a selection and training format designed to weed out people who simply don't belong. And that usually ends up meaning the vast majority. The recruits, who must be under twenty-eight years old, are eliminated by the instructors as they slowly but inexorably increase the pressure on each candidate to make sure only the strongest survive. So tough is the regime that sometimes it results in nearly every candidate failing. It is a deliberately brutal experience that is not pretty to watch and pushes its victims to the limits of their physical and emotional endurance, and beyond. Injuries are routine and deaths in training sometimes occur.

SEAL instructors are adamant that it isn't brute strength that is required to survive the training course but an attitude of mind, a strength of character, a kind of internal motivation rather than pure physical strength. They maintain that if a recruit has the mental strength to work through the discomfort, the fatigue and the humiliation, he can usually develop the physical strength necessary. The head is the hard part.

Although the BUDS training takes half a year, it is only one part of the learning process of becoming a full qualified SEAL. For a start, the medical is daunting. Eyesight must be 20/40 in one eye and 20/20 in the other, with no colour blindness. More importantly, the recruit must pass the physical fitness tests. These include swimming five hundred yards in less than twelve and a half minutes; resting for ten minutes and then carrying out at least forty-two press-ups in two minutes, fifty sit-ups in two minutes, eight pull-ups, and then running a mile in boots and pants in eleven and a half minutes – all in less than one hour. In addition, high scores must be achieved on military written tests.

New arrivals spend the first seven weeks enduring a programme of indoctrination and physical preconditioning, with long hours of classes, running, swimming, sit-ups, press-ups and gymnastics before they start the three-phase BUDS course.

Phase One lasts nine weeks and features almost non-stop swimming, running and tackling assault courses. Every trainee is required to put maximum effort into every test on every occasion. The minimum score is raised after every discipline is completed and every trainee has to better his previous score. As the instructors never tire of telling the recruits, 'The only easy day was yesterday.'

One of the toughest challenges is the obstacle course, which looks like a giant sandpit with telephone poles assembled in a wide variety of structures. Once a week for an hour or two the recruits are ordered to dash around the pre-ordained circuit, which demands much running, jumping, net-climbing, crawling under barbed wire, hopping from pole to pole and hauling the body up and over high wooden walls. And every week every recruit must better his previous time.

The recruits spend much of their day in the water, generally the Pacific Ocean, which for most of the year is cold, sometimes so cold that recruits suffer hypothermia. Then the recruit must stand in the chill wind on the beach, shivering from the cold. Some have died.

And after five weeks of that comes what is aptly named 'Hell Week'. This begins shortly after midnight, when instructors wake up recruits with their own alarm, firing M60 blanks and artillery simulators, creating a deafening noise designed to temporarily traumatise the recruits. For the following six days the recruits are allowed twenty minutes' sleep a day, and have to move from one discipline to another all the time. They will run on the beach, carry out boat drills, do PT, swim and then start again.

As the US Special Forces handbook states:

One day melts into another without rest. There is no alternative but to tough it out, drive through the fatigue and keep doing what they tell you to do. It is a test of mental toughness as much as the powers of physical endurance. After four days or so people start to

hallucinate. And some people start to quit. Hell Week is the most important week of the whole BUDS training programme, a physical and mental challenge that is intended to put the trainees under stress that is supposed to approach that of actual combat.

Phase Two teaches recruits everything there is to know about diving operations, including scuba diving, closed-circuit re-breathing systems and the physiology of diving. They learn how to deal with equipment failures, lost regulators and the hazards of nitrogen narcosis. In Phase Three the recruits learn the art of underwater demolition, land navigation, explosives, small unit tactics, abseiling and the use of SEAL weapons.

At any time during the six-month course any recruit can quit voluntarily, but anyone who does so must face great humiliation. The recruit stands on green-painted frog footprints in the main gymnastics area, holds the lanyard of a ship's bell and rings the bell three times. He then turns, places his green helmet liner on the pile of liners of other 'quitters' and marches off.

The BUDS training schedule has been criticised for unnecessary extremes; the injuries, the high levels of stress and the humiliation are all far more than anyone in civilian life ever endures. And yet the recruits are pushed to these extremes so that when they actually face action they have the mental and physical strengths to survive.

If and when the recruit passes those tough tests he moves to Fort Benning, in Georgia, for the five-week Basic Airborne Course, in order to become a fully qualified paratrooper. Compared to BUDS, the parachute course is a doddle.

And when this is completed the recruit is put on probation for six months, during which time he can still be failed if his team mates find he isn't up to scratch. At the end of this period a few young men will be entitled to pin the big, gold symbol of Naval Special Warfare on their chests and finally call themselves SEALs.

Despite the explosion of Special Forces soldiers the world over, there is still the big question which soldiers everywhere talk about, argue about and want an answer to: 'What *makes* a Special Forces soldier?'

The answers come thick and fast from various sources. Robin Neillands, in his book *In The Combat Zone*, sums up the view of many when he says:

In my opinion there are some folks who are just instinctive warriors. Combine that instinct with the pioneering spirit ... and the titillating rush of a little or a lot of danger and you get the special soldier. Maybe it is someone who just wants to be different, for whom the commonplace, the safe and the banal is simply not enough. Someone who wants to look back at their life and believe that they did something to change the course of history.

The Special Forces calls a category of its operations DAs – Direct Actions – and maybe therein lies the clue. Forget all the complications of bureaucratic soldiering, just point us at the bad guys and let 'er rip. Those of us who are the breed know one another ... a few words of conversation and the 'duffers' are separated from the long-ball hitters. It doesn't take a ton of words from a Special Soldier's mouth to let another Special Soldier know that they are warrior kin.

Finally, many senior NCOs involved in training would-be Special Forces soldiers have identified another important reason why young men, despite knowing that the rejection rate is very high and the course unbelievably demanding, want to join these elite forces. They believe that what initially motivates these young men – mostly in their mid-to late twenties – to join is simply an addiction to adrenalin. Such a man is seeking to gain respect and to prove himself among young men he already respects, to achieve an ambition which the world respects and admires, and to prove to himself that he is one of the few, the select, the best.

THE WORLD'S LEADING SPECIAL FORCES

United States

US Army: 75th Ranger Regiment; 160th Special Operations Aviation Regiment (Airborne); US Army Special Operations Command (USASOC)

US Navy: Sea, Air, Land Forces (SEALs)

US Air Force: Special Operations Squadrons

Joint Special Operations Command (JSOC)

United Kingdom

16 Air Assault Brigade, including HQ and Signal Squadron; Pathfinder Platoon, Army Air Corps; 7th Para RHA; 21st Defence Battery, RA; 9th Parachute Squadron RE; RAF Support Helicopter Force; 47 Air Despatch Squadron

Parachute Regiment

Royal Marines

Special Air Service (SAS)

Special Boat Service (SBS)

Gurkha Regiment

France

Commandement des opérations spéciales (COS)

1 Régiment parachutiste d'infanterie de marine (I RPIMa)

Détachement aérien des opérations spéciales (DAOS)

Groupement spécial autonome (GSA)

Commandement des fusiliers marins commandos (COFUSCO)

Commando parachutiste de l'air no. 10 (CPA 10)

Escadrille des hélicoptères spéciaux (EHS)

Division des opérations spéciales (DOS)

Germany

Gebirgsjäger (Mountain Infantry Division)
Fallschirmjäger (Parachute Infantry Brigade)
Kommando Spezialkrafte (Commando Brigade)

Russia

Spetsnaz (Spetsialnoje Naznachenie – forces of Special Designation)
Razvedchiki (Long-range and airborne operations)
Naval Spetsnaz (Amphibious reconnaissance and operations)
Morskaya Pekhota (Marine Commandos)

Belgium

Paratroop Commando Brigade
1st Battalion (Paratroopers and commandos)
Frogmen (Trained on lines of Britain's Special Boat Service)

Italy

Alpini Brigades (Mountain fighters)
Raggruppamento Anfibio San Marco (Rapid intervention force)
Commando Raggruppamento Subacqui ed Incursori (Comsubin)
(Assault divers)
Gruppo Operativo Incursori (GOI) (Includes combat divers, parachutists, helicopter-borne commandos)

Netherlands

Nederlands Korps Mariniers (KNKM) (Special amphibious force)
1st Marine Battalion and 7 Troop SBS (Amphibious force capable of operating worldwide)
Bijzondere Bijstands Eenheid (BBE) (Counter-terrorist unit)
Korps Commandotroepen (KCT) (Commandos)
108 Special Forces Company (Commando special operations)
11th Air Mobile Brigade (Rapid-deployment airborne special forces)

Spain

Fuerzo de Acción Rápido (FAR) (Rapid-reaction force)
Brigada Paracaidista (BRIPAC) (Parachute Brigade)
Unidad Especial de Buceadores de Combate (UEBC) (Water-borne
special forces)
Tercio de Armada (TEAR) (A Marines brigade)
1st Airborne Bandera (Free-fall parachutists)
Unidades de Operaciones Especiales (UOE) (Three separate forces
involved in anti-terrorism, anti-guerilla warfare and
commando operations)

Greece

Special Forces Directorate
1 Parachute Regiment
1 Marine Brigade
1 Special Operations Command
1 Ranger Regiment

Turkey

1, 2 and 3 Airborne Platoons of the Guards Regiment (All three platoons
are battalion-strength commandos, marines and paratroopers)
1st Commando Brigade (Underwater defence)
2nd Commando Brigade (Underwater attack)
3rd Commando Brigade (Long-range amphibious intelligence-gathering
and sabotage)

Israel

202nd Parachute Regiment (Spearheads every war against Arab nations)
Sayaret (Deep-penetration raids, anti-commando missions
and reconnaissance)
Golani Brigade (Infantry brigade – Israel's premier elite unit)
Parachute Reconnaissance
Sayaret Matkal (GHQ Reconnaissance, unofficially called Unit 269 –

NICHOLAS DAVIES

special commando force under GHQ Intelligence Corps)
Israeli Border Guard (Anti-terrorist and special operations)
Yamam (Counter-terrorist and hostage-rescue unit)
Yamas (Highly secret unit which conducts undercover operations against
Palestinian fighters in the West Bank and Gaza Strip)
The Naval Commando (13th Flotilla) (Based on Britain's
Special Boat Service)

Egypt

A-Saiqa (Lightning) Commandos
130th Marine Brigade
182nd Paratrooper Brigade
Force 777 (counter-terrorist unit)

Syria

Al-Wahdat al-Khassa (Special Units – elite forces similar to Britain's
Special Air Service)
14th Special Forces Division (four regiments)
Saraya al Difa (paratroopers)

North Atlantic Treaty Organisation (NATO)

Rapid-reaction special forces are available at all times to the Supreme
Allied Commander Europe
Immediate-reaction forces are on three-day deployment
Rapid-reaction forces are on seven-day deployment
Elite Commando, Paratroop and Special Forces forces from all
NATO countries

APPENDIX 2

SPECIAL FORCES WEAPONS

Revolvers and Pistols

Smith & Wesson Model 624
Made in USA. Double-action personal defence revolver. Operation:
manual, single shot only. Calibre: .44. Length: 9.13in. Weight: 36oz.
Magazine: six-chamber rotating cylinder. Muzzle velocity: 1470ft/sec.

FN Browning GP-35

Made in Belgium. Semi-automatic pistol. Operation: recoil, semi-
automatic only. Calibre: 9mm. Cartridge: 9mm x 19mm, rimless. Length:
7.75in. Weight: 35oz. Barrel: 4.72in. Magazine: thirteen-round,
detachable. Muzzle velocity: 1148 ft/sec.

CZ-85

Made in Czech Republic. Semi-automatic pistol. Operation: recoil, semi-
automatic only. Calibre: 9mm. Cartridge: 9mm x 19mm, rimless. Length:
8in. Weight: 35oz. Barrel: 4.72in. Magazine: fifteen-round, detachable.
Muzzle velocity: 1300 ft/sec.

Walther P88

Made in Germany. Semi-automatic pistol. Operation: recoil, semi-
automatic only. Calibre: 9mm. Cartridge: 9mm x 19mm, rimless. Length:
7.13in. Weight: 29oz. Barrel: 3.82in. Magazine: fourteen-round,
detachable. Muzzle velocity: 1148 ft/sec.

Beretta Model 92F Compact

Made in Italy. Semi-automatic pistol. Operation: recoil, semi-automatic
only. Calibre: 9mm. Cartridge: 9mm x 19mm, rimless. Length: 7.76in.
Weight: 32oz. Barrel: 4.29in. Magazine: thirteen-round, detachable.
Muzzle velocity: 1280 ft/sec.

Desert Eagle

Made in Israel. Semi-automatic pistol. Operation: gas, semi-automatic
only. Calibre: 0.357. Cartridge: 0.357 Magnum, rimmed. Length: 10.24 in.
Weight: 60oz. Barrel: 5.91in. Magazine: nine-round, detachable. Muzzle
velocity: 1430 ft/sec.

Ruger P95

Made in USA. Semi-automatic pistol. Operation: recoil, semi-automatic only. Calibre: 9mm. Cartridge: 9mm x 19mm, rimless. Length: 7.3in. Weight: 27oz. Barrel: 3.9in. Magazine: ten-round, detachable. Muzzle velocity: 1148 ft/sec.

Glock 17

Made in Austria. Semi-automatic pistol. Operation: recoil, semi-automatic only. Calibre: 9mm. Cartridge: 9mm x 19mm, rimless. Length: 7.32in. Weight: 22oz. Barrel: 4.50in. Magazine: seventeen-round, detachable. Muzzle velocity: 1263 ft/sec.

FN Five-seveN

Made in Belgium. Semi-automatic pistol. Operation: delayed blowback. Calibre: 5.7mm. Cartridge: 5.7mm x 28mm P90. Length: 8.19in. Weight: 21.8oz. Barrel: 4.82in. Magazine: twenty-round, detachable. Muzzle velocity: 2133 ft/sec.

Heckler & Koch USP

Made in Germany. Semi-automatic pistol. Operation: recoil, semi-automatic only. Calibre: 9mm. Cartridge: 9mm x 19mm, rimless. Length: 7.64in. Weight: 25.4oz. Barrel: 4.25in. Magazine: fifteen-round, detachable. Muzzle velocity: 1148 ft/sec.

Shotguns

Franchi SPAS-12

Made in Italy. Auto-loading/slide-action shotgun. Operation: gas or manual, single shot only. Calibre: 12-bore. Cartridge: 12mm x 70mm, rimmed. Length: 36.61in. Weight: 9.15lb. Barrel: 18in. Magazine: six-round tube beneath barrel. Muzzle velocity: N/A.

Mossberg Model 500 Persuader

Slide-action shotgun. Operation: manual, single shot only. Calibre: 12-bore. Cartridge: 12mm x 70mm, rimmed. Length: 38.5in. Weight: 6.75lb. Barrel: 18.5in, smooth-bore. Magazine: six-round tube beneath barrel. Muzzle velocity: N/A.

Infantry Rifles

AK74M
Made in Russia. Combat rifle; current version of the legendary AK47.
Operation: gas, selective fire. Calibre: 5.45mm. Cartridge: 5.45mm x
39mm M74, rimless. Length: 37.64in. Weight: 10.69lb. Barrel: 16.34in.
Magazine: thirty-round, detachable. Muzzle velocity: 2953 ft/sec.

M16A2

Made in USA. Combat rifle. Operation: gas, selective fire. Calibre:
5.56mm. Cartridge: 5.56mm x 45mm, rimless. Length: 39.37in. Weight:
7.50lb. Barrel: 20in. Magazine: thirty-round, detachable. Muzzle velocity:
3110 ft/sec.

Beretta AR70/90

Made in Italy. Operation: gas, selective fire. Calibre: 5.56mm. Cartridge:
5.56mm x 45mm, rimless. Length: 39.17in. Weight: 8.69lb. Barrel:
17.72in. Magazine: thirty-round, detachable. Muzzle velocity: 3052 ft/sec.

Light Automatic Weapons

L2A3 Sterling
Made in Britain. Sub-machine gun. Operation: blowback, automatic or
selective fire. Calibre: 9mm. Cartridge: 9mm x 19mm, rimless. Length:
27.17in. Weight: 6.0lb. Barrel: 7.8in. Magazine: thirty-four-round,
detachable. Muzzle velocity: 1280 ft/sec.

Beretta M12S

Sub-machine gun. Operation: blowback, selective fire. Calibre: 9mm.
Cartridge: 9mm x 19mm, rimless. Length: 25.40in. Weight: 6.56lb.
Barrel: 7.9in. Magazine: twenty-thirty-two-or forty-round, detachable.
Muzzle velocity: 1250 ft/sec.

Heckler & Koch MP5A4

Sub-machine gun. Operation: delayed blowback, selective fire. Calibre:
9mm. Cartridge: 9mm x 19mm, rimless. Length: 26.77in. Weight: 5.60lb.
Barrel: 8.86in. Magazine: fifteen-or thirty-round, detachable. Muzzle
velocity: 1313ft/sec.

Light Machine Guns

FN Minimi

Made in Belgium. Light support weapon. Operation: gas, selective fire.
Calibre: 5.56mm. Cartridge: 5.56mm x 45mm, rimless. Length: 40.94in.
Weight: 15.06lb. Barrel: 18.35in. Feed: Two hundred-round metal-link
belt or thirty-round, detachable. Muzzle velocity: 3035 ft/sec.

Negev

Made in Israel. Light support weapon. Operation: gas, selective fire.
Calibre: 5.56mm. Cartridge: 5.56mm x 45mm, rimless. Length: 40.16in.
Weight: 15.87lb. Barrel: 18.11in. Feed: metal-link belt or detachable box.
Rate of fire: 650–850 or 750–950 rnds/min. Muzzle velocity: 3117 ft/sec.

Colt LMG M16A2

Made in USA. Light support weapon. Operation: gas, automatic fire only.
Calibre: 5.56mm. Cartridge: 5.56mm x 45mm, rimless. Length: 39.60in.
Weight: 12.75lb. Barrel: 20in. Magazine: thirty-round, detachable. Rate
of fire: 600–750 rnds/min. Muzzle velocity: 3250 ft/sec.

SELECT BIBLIOGRAPHY

Adams, James, *The Financing of Terror – Behind the PLO, IRA, Red Brigades and M19 Stand the Paymasters*, Simon & Schuster, 1986

Adkin, Mark, *Urgent Fury – The Battle for Grenada*, Leo Cooper, 1989

Arostegui, Martin C., *Twilight Warriors – Inside the World's Special Forces*, Bloomsbury Publishing, 1995

Becker, Jillian, *Hitler's Children – The Story of the Baader-Meinhof Terrorist Gang*, Pickwick Books, 1989

Betser, Colonel Muki, and Rosenberg, Robert, *Secret Soldier*, Simon & Schuster, 1996

Billiere, General Sir Peter de la, *Looking for Trouble – SAS to Gulf Command*, HarperCollins, 1994

Bowden, Mark, *Black Hawk Down*, Bantam Press, 1999

Clutterbuck, Richard, *Living with Terrorism*, Arlington House, 1975

Darwish, Adel, and Alexander, Gregory, *Unholy Babylon – The Secret History of Saddam's War*, Victor Gollancz, 1991

Follian, John, *Jackal – The Secret Wars of Carlos the Jackal*, Orion, 1999

Geraghty, Tony, *Who Dares Wins – The Special Air Service 1950 to the Gulf War*, Little, Brown, 1992

Harclerode, Peter, *Secret Soldiers*, Cassell, 2000

Harnden, Toby, *Bandit Country – The IRA and South Armagh*, Hodder & Stoughton, 1999

Hastings, Max, *Yoni – Hero of Entebbe*, Weidenfeld & Nicolson, 1979

Hunter, Robin, *True Stories of the Foreign Legion*, Virgin, 1997

Hunter, Robin, *True Stories of the SBS*, Virgin, 1998

Jeapes, Major-General Tony, *SAS Secret War*, HarperCollins, 1996

Kennedy, Michael Paul, *Soldier I SAS*, Bloomsbury Publishing, 1990

Landau, Alan M., and Frieda W., *U.S. Special Forces*, MBI Publishing, 1999

Lucas, James, *Kommando*, Arms and Armour Press, 1985

Neillands, Robin, *In The Combat Zone*, Weidenfeld & Nicolson, 1997

Owen, David Lloyd, *Providence Their Guide*, Harrap, 1980

Parker, John, *SBS – The Inside Story of the Special Boat Service*, Headline, 1997

Seale, Patrick, *Abu Nidal: A Gun for Hire*, Random House, 1992

Seymour, William, *British Special Forces*, Sidgwick & Jackson, 1985

Sterling, Claire, *The Terror Network – The Secret War of International Terrorism*, Holt, Rhinehart & Winston, 1981

Tophoven, Rolf, *GSG-9 – German Response to Terrorism*, Bernard & Graefe Verlag, 1984

White, Terry, *Fighting Skills of the SAS and Special Forces*, Century Random Press, 1993

INDEX

INDEX

East Africa 20
East Tyrone Brigade (Provisional IRA) 150
EC-130 transport plane 169, 170
Egypt 117, 168, 183, 185
Eighth Army (Britain) 20–1, 28
El Alamein 21
Elgin Air Force Base 254
Endurance, HMS 133, 134
Entebbe 81, 84, 107, 109–17
ETA (Euzkadi ta Azkatasuna: Basque Nation and Liberty) 81
Eversmann, Staff Sergeant Matt 207, 209, 210, 212, 213, 215, 216

F-104 Starfighter fighter aircraft 99
Falkland Islands 123, 124, 125, 127, 128, 130, 131, 132, 135, 136, 138, 141, 251
    East Falkland 138
    Governor-General of 127, 128, 129
    Pebble Island 138, 139
    West Falkland 138
Falklands War (1982) 123–41
Farrell, Maraid 159, 160, 161, 162, 163, 164
FBI 83
FN rifle 150
Foreign Office 48, 129, 131
Fort Benning 252, 257
Fort Rupert 172, 175
France 110, 179, 183, 245
Frank, Chief Warrant Officer Ray 222, 223, 226
Frankfurt 85, 167, 168
Free Fall Troop, SAS 250
French Foreign Legion 55–75, 178, 184
    formation of 67
    in Algeria 68–9
    in Indo-China 58–67
    in Libya 178–9
    in Zaire 69–74
    selection for 57–8
    training for 58
Fuller, Chief Warrant Officer Gary 231

Gadaffi, Col Muammar 81, 178, 181
Galtieri, General Leopoldo, President of Argentina 129
Garrison, General William 212, 214, 215, 223, 231
Gazelle helicopter 138
Geneva Convention 32, 123
Gennep bridge 15–16
Geraghty, Tony 37
Germany 245
Germany, West 83, 84, 179
Giap, General Vo Nguyen 58–60, 65, 66
Gibraltar 159, 160, 161, 162, 164
Glamorgan, HMS 139
Goffena, Chief Warrant Officer Mike 224, 225, 226, 227, 228
Golan Heights 111
Gorbachev, Mikhail, President of the Russian Republic 244
Gordon, Master Sergeant Gary 225, 226, 227, 228, 229
Grabert, Lieutenant Klaus 16, 17
Great War (World War One) (1914–18) 12, 16
Green Berets (USA) 177, 184, 197
Grenada 171–7
Grenadian Defence Force 173
Grentzschutzgruppe-9 (GSG-9) (West Germany) 83–4, 85, 87–91
Grozny 103
GRU 240
Grytviken 124, 134, 135
Guerrico 124, 125, 126
guerrilla warfare 33–45
guerrillas 12
Gulf War (1990) 183–205
Gurkhas 34, 35, 259

Habash, George 85
Habr Gidr 208, 210
Haddad, Wadi 80, 85
Hamshari, Mahmoud 119

Harb, Wabil 86, 90, 91
Harclerode, Peter 47, 170
Harrier jump jet 137
Hay, Lieutenant Colonel Tony 28
Headquarters Mobile Support Unit (HMSU), RUC 160
Heathrow airport 47
Heckler & Koch machine pistol 192
Heckler & Koch MP5 sub-machine gun, 50, 51, 162, 200, 204, 265
Heckler & Koch rifle 150
Heidelberg 85
'Hell Week' 256–7
Hercules C-130 transport plane 70, 111, 112,113, 115, 116, 117
Herse, Abdi Yusef 210
Hitler, Adolf 12, 14, 15, 19, 32
Ho Chi Minh 58, 59
Humvee 208, 212, 215, 216, 217, 218, 220, 221, 222, 223, 224, 233
Hussein, Saddam, head of state of Iraq 183, 184, 185, 186, 187, 198, 201

In the Combat Zone 141, 158–9, 258
Indo-China 58–67
Indonesia 95, 96
Intelligence 13, 21, 45, 48, 81, 82, 134, 135, 159, 161, 176, 208
IRA (Irish Republican Army), Provisional 81, 147–50, 157–64
IRA, Official 157
Iran 165, 166, 167, 168, 169, 172
    Shah of 80, 165
    war with Iraq 184
Iraq 81, 87, 184, 185, 186, 187, 189, 201
    war with Iran 184
Israel 78, 80, 83, 84, 109, 110, 185

Jabal Akhdar 36–40
Jewish Night Squads 111
Joint Operations Center (JOC) 214
JS tank 23–5
jungle warfare 34–5

Kabul 237, 240
Kalashnikov AK47 combat rifle 40, 41, 148, 208, 210, 211, 216, 222, 231, 232, 239, 242, 265
Kandahar 237
Kealy, Captain Mike 42, 43, 44
Kelly, Patrick 150
KGB 238, 239, 243, 244
Khaled, ibn Abdul Aziz, King of Saudi Arabia 80
Khan, Genghis 9
Khomeini, Ayatollah 165, 166, 170
King Edward Point 133, 134
Kinshasa 70
Kolwezi 69–72, 74
Kowalewski, Private Richard ('Alphabet') 219, 220
Kröcher-Tiedemann, Gabrielle 107, 108, 113
Kubeisy, Dr Bassel Rauf 119
Kuwait 87, 108, 183, 184, 186, 187
Kuwait City 183

'Leathernecks' see Nucelo Operativo
    Centrale di Sicurezza
Lavasani, Abbas 48, 49
Lawrence, T.E. (Lawrence of Arabia) 13
Lee Enfield .303 rifle 41
Leoni, Raul, President of Venezuela 77
Libya 21, 22, 109, 179
Libya 75, 80, 81, 178, 179, 181
Life Guards (Britain) 38
Lillehammer 120
Lloyd Owen, Major General David 28,29
Lock, PC Trevor 46–8, 51
Lockheed C-141 Starlifter transport plane 167
London 45–53, 77, 78, 129, 135
Long Range Desert Group (Britain) 21, 27–30
Loughgall 147
Luftwaffe 28, 32

270